Voyage to the End of the World

with Tales from the Great Ice Barrier

David Burke

Published by the University Press of Colorado
5589 Arapahoe Avenue, Suite 206C
Boulder, Colorado 80303

Printed in the United States of America

The University Press of Colorado is a proud member of the Association of American University Presses.

The University Press of Colorado is a cooperative publishing enterprise supported, in part, by Adams State College, Colorado State University, Fort Lewis College, Mesa State College, Metropolitan State College of Denver, University of Colorado, University of Northern Colorado, and Western State College of Colorado.

∞ The paper used in this publication meets the minimum requirements of the American National Standard for Information Sciences—Permanence of Paper for Printed Library Materials. ANSI Z39.48-1992

Library of Congress Cataloging-in-Publication Data

Burke, David, 1927–
Voyage to the end of the world : with tales from the great ice barrier / David Burke.
p. cm.
Originally published: Annadale, N.S.W., Australia ; EnviroBook, 2002.
Includes bibliographical references and index.
ISBN 0-87081-771-X (alk. paper)
1. Kapitan Khlebnikov (Icebreaker) 2. Voyages and travels. 3. Antarctica. 4. Antarctic Ocean. I. Title.

G530.K19B87 2004
919.8'904—dc22

2004055494

14 13 12 11 10 09 08 07 06 05 10 9 8 7 6 5 4 3 2 1

Amid the Ross Sea floes, the powerful Russian ship Kapitan Khlebnikov *seems to be in its natural element—breaking ice.*

THE BARRIER BERG

Breasting the slow-heaving, limitless swell,
Child of the snow and the far-reaching tides,
Proudly I swing, sheer-riven, clean-cornered,
Dipping awash my immaculate sides.

Drift of the ocean beneath me, compelling,
Drives me athwart half the winds of the world,
Soft tepid seas, embracing, encave me,
Foam-lathered billows beset me upcurled.

Broken and tilted, sun-wasted, sea-drunken,
Once more to the shores of the Southland returned,
Aground on a shoal I await the last mercy,
Swift death and complete in the cold sea interned.

—*F. Debenham*[1]
McMurdo Sound, 1911

Contents

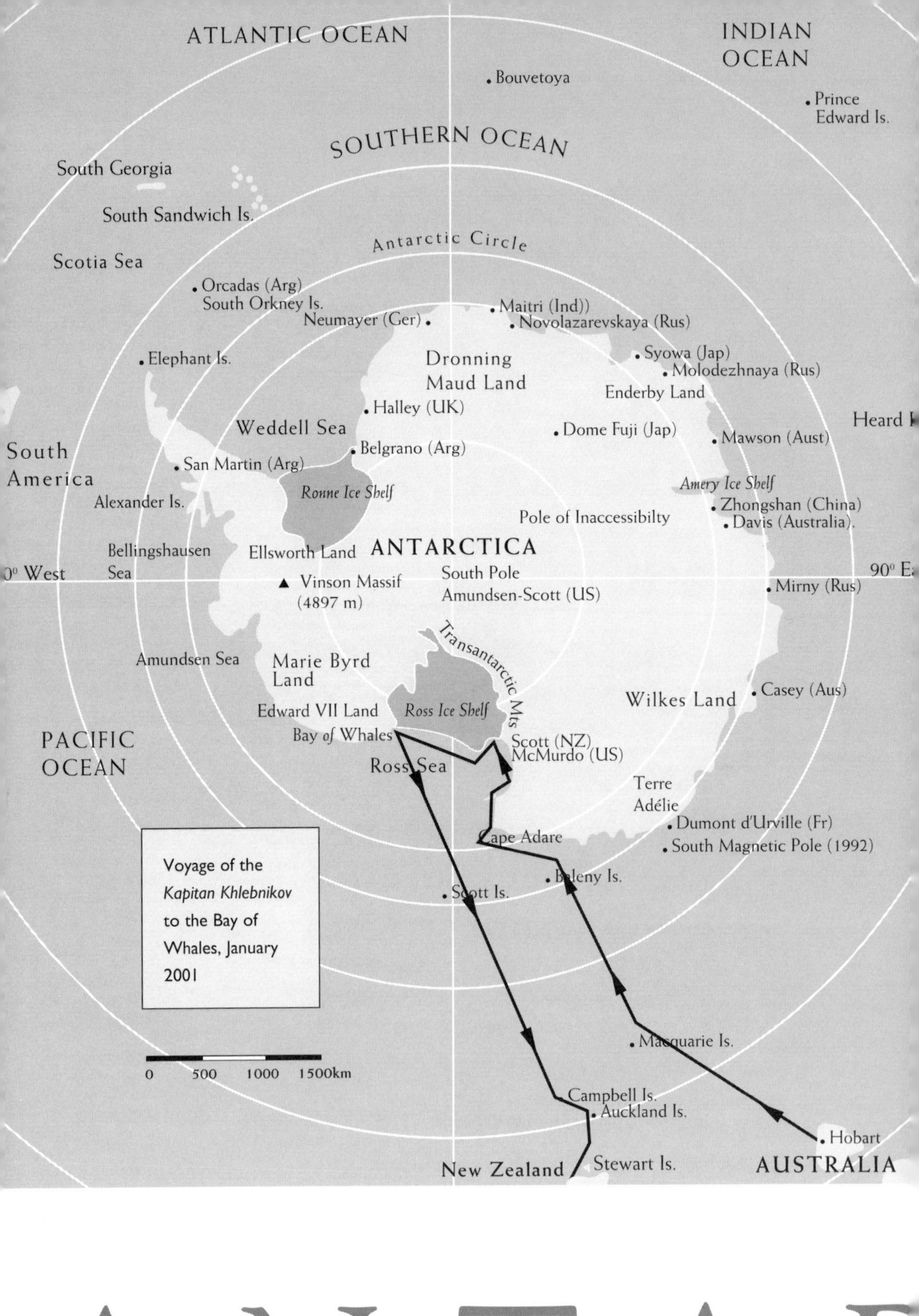

ATLANTIC OCEAN
INDIAN OCEAN
Bouvetoya
Prince Edward Is.
SOUTHERN OCEAN
South Georgia
South Sandwich Is.
Antarctic Circle
Scotia Sea
Orcadas (Arg)
South Orkney Is.
Neumayer (Ger)
Maitri (Ind))
Novolazarevskaya (Rus)
Elephant Is.
Dronning Maud Land
Syowa (Jap)
Molodezhnaya (Rus)
Enderby Land
Halley (UK)
Weddell Sea
Dome Fuji (Jap)
Mawson (Aust)
Heard
South America
San Martin (Arg)
Belgrano (Arg)
Amery Ice Shelf
Alexander Is.
Ronne Ice Shelf
Zhongshan (China)
Davis (Australia).
Pole of Inaccessibilty
Bellingshausen Sea
Ellsworth Land
ANTARCTICA
West
90°
Vinson Massif (4897 m)
South Pole
Amundsen-Scott (US)
Mirny (Rus)
Transantarctic Mts
Amundsen Sea
Marie Byrd Land
Wilkes Land
Casey (Aus)
Edward VII Land
Ross Ice Shelf
Bay of Whales
PACIFIC OCEAN
Scott (NZ)
McMurdo (US)
Ross Sea
Terre Adélie
Dumont d'Urville (Fr)
Cape Adare
South Magnetic Pole (1992)
Balleny Is.
Scott Is.
Voyage of the Kapitan Khlebnikov to the Bay of Whales, January 2001
0 500 1000 1500km
Macquarie Is.
Campbell Is.
Auckland Is.
Hobart
New Zealand
Stewart Is.
AUSTRALIA

A N T A

Foreword

The author first visited Antarctica in 1958—a year in which the polar continent seemed barely to have stirred from the heroic age of Amundsen, Scott, Shackleton, and Mawson. Ice-encrusted and silent, the wooden huts of the early British explorers still nestled around the shores of McMurdo Sound; gaunt crosses still lifted above the icy slopes of Ross Island, in memory of lost men whose bodies lie beneath the snow.

For a young newspaperman, it was an exciting year to be sent South to report the new ice age of science and machine. A Commonwealth expedition led by Sir Vivian Fuchs was crossing the continent; air transport was penetrating to Antarctica's far corners, and men were living on that cold hell known as the polar plateau.

Forty-three years later the author returned on his sixth mission South—this time to participate (but with considerably improved comfort) in a unique voyage to the bottom of the world, an attempt by a Russian icebreaker to reach the "farthest south" a ship had ever sailed.

Much has changed in those years between. The South Pole has become a permanent observatory, the American base at McMurdo houses upwards of 1,000 people every summer season, solo adventurers pit themselves against crevasse and blizzard, and tourism is on the march.

Sometimes I count myself fortunate to have been an observer at the coming to Antarctica of this modern age. To have stood at Scott Base and witnessed, as if in an echo of the heroic past, the end of an epic march across the continent—the Commonwealth expedition's realization of Shackleton's frustrated dream. To have participated in the pioneering intercontinental flight between Australia and the Pole, with all its unbidden adventure. To have gazed, where few eyes have been privileged, at glistening peaks, rank upon rank, glacier after glacier, of the Trans-Antarctic Mountains and beneath them, to trudge the snows of the Ross Ice Shelf for the making of a tiny weather station at the foot of the mighty Beardmore.

To have lived in a tunnel cut beneath the ice at lofty Byrd Station on the Rockefeller Plateau, and along the Antarctic Peninsula to watch (a trifle too close for comfort) the death throes of an immense iceberg. To ride the earliest convoy attempting to crash a path through the encircling winter pack, to gaze down on the wild ice of untrodden Oates Land and finally to stand on the bridge of a Russian 'breaker, thrusting through the Ross Sea, bows fixed on a point in the Great Ice Barrier in a quest for Farthest South. Such memories are entries in a diary of one on whom Antarctica has cast its spell. But nothing defined as "progress" dares touch that most remote and historic of Antarctic locations—at the far end of the Great Ice Barrier—the Bay of Whales, where the big 'bergs are born.

BRISBANE'S LARGEST AUDITED DAILY CIRCULATION: Last week 223,303

The Courier-Mail

BRISBANE MONDAY MARCH 3 1958

[Printed and published by Queensland Newspapers Ltd., of 268 Queen Street, Brisbane, at that add

Telephone FA0111

FANTASTIC SCEN CLIMAX THE CROS OF ANT

SOUTH POLAR trek hero, Britain's Dr. Vivian Fuchs — knighted by the Queen.

From DAVID BURKE, the only Australi

SCOTT BASE (Antarctica).—The a land crossing of the A here at 1.47 p.m. yesterday, le soon after was knighted by th

In brilliant sunshine the ex coloured Snocats roared into S horns sounding, and the bewhi waving from their windows.

It was probably the most fantas arctic has known as the welcomi than 150 men – Americans from British and Commonwealth me —gathered to cheer them in.

Hissing flares and brilliant red sent up, streamers were let loose, an formed "philharmonic band" burst

History was made as the bearded fig down from his Sno-cat "Rock 'n' Roll, through the guard of honour—which al tractors, weasels, carriers, and sledges.

He and his 11-man expedition, with members from Britain, New Zealand, Australia, and South Africa, had completed the 2100-mile crossing of Antarctica from the Weddell Sea to the Ross Sea by way of the South Pole, in exactly 99 days.

Waiting for Sir Vivian in the tiny Scott Base post office was a cablegram message from the Queen saying: "Well done."

Home

WELLINGTO (A.A.P.). future plans, Fuchs said y want to get land, where waiting to s

He would had plans expedition, "One neve one expedi

Sir Edm "I am an garden is

American icebreakers during Operation Deep Freeze enter McMurdo Sound, where Antarctica's active volcano, the 12,448-foot (3,795-meter) Mount Erebus, lifts into the sky above Ross Island.

Introduction

The Bay of Whales lies close to the eastern extremity of the Ross Ice Shelf in the region of 78° South latitude and 164–165° West longitude. This is as close as a ship can get to the heart of Antarctica. The Ross Ice Shelf—in reality an immense floating glacier—and the cliffs of the Great Ice Barrier that enclose it are, in the words of an old explorer, "one of this planet's most awesome spectacles." The Ice Shelf—indeed, "shelf" seems a hopelessly inadequate term for an ice table that exceeds in area that of the state of Montana—drains the glacial outpourings of the Trans-Antarctic Mountains; for about 450 miles (720 km) the Barrier front presents a line of superb and sheer white cliffs rising some 130 feet (40 m) above the Ross Sea.

At the far western end of the Barrier lies Ross Island dominated by the 12,448-foot (3,795-meter) active volcano, Mt. Erebus, and below it McMurdo Sound, first sighted by Captain James Clark Ross RN in 1841. New Zealand claims the Ross Sea region under the name of the Ross Dependency, while across from McMurdo, bordering the

Trans-Antarctic Mountains, the vast Australian claim to Antarctica begins. Beyond the Ross Dependency at the eastern end of the Barrier lies the terra nullius of Marie Byrd Land, named by Richard E. Byrd for his wife, in making an American claim that the U.S. government has preferred not to exercise. The Bay of Whales, one of Ross's discoveries in his voyages of 1840–1843 with his ships *Erebus* and *Terror,* is an icy repository of Antarctic history. Indeed for explorers of the heroic age and afterwards it held the key to winning the South Pole. From hereabouts in 1902, Captain Robert Falcon Scott made the first balloon ascent of Antarctica. Shackleton sailed by in 1908 and gave the Bay its name, but fearing the ice might collapse decided not to stay. Three years later Amundsen arrived and within another year had planted the Norwegian flag at the South Pole. Commander Richard E. Byrd promoted Antarctica's modern era in the pioneering flight of 1929 from the Bay to the South Pole. The American millionaire, Lincoln Ellsworth, completed the first flight across a sector of Antarctica at the Bay in 1935, and soon afterwards an Australian party arrived to rescue him in one of the more bizarre episodes of polar history. The U.S. Navy descended upon the Bay in 1947 with the huge "war games" force of Operation Highjump to be followed within another eight years by a second invasion, but this time it was Operation Deep Freeze and the mission was science, international co-operation, and peace.

Today not a trace of the polar dramas enacted upon these icy cliffs remains. The Bay of Whales and the enclosing Barrier are wholly indifferent to man's ambitions. Impermanence and change are their only constant. The sound of barking huskies, the rattle of ships' winches and the revving of aircraft engines no longer echo against the surrounding cliffs. The icy ramparts where famous expeditions and their leaders once made headlines have drifted to a watery tomb beneath the Ross Sea. A voyage to the end of the world will recall the Barrier ghosts and the dramas of its ever-changing Bay.

Where Big 'Bergs are Born

In March 2000, satellite images revealed the calving of possibly the largest iceberg the world had ever known. Numbered "B15" for tracking purposes, the immense 'berg broke towards the eastern end of the Ross Ice Shelf and extended its fracture westwards for a distance of some 190 miles (340 km). This was the natural wonder that the travellers of *Kapitan Khlebnikov* had come to see. With a width of about twenty-five miles (forty km), the 'berg was calculated to cover 4,250 square miles (10,500 km^2), an area

A Vladivostok Air helicopter has landed photographer Catherine on top of the giant iceberg B15D.

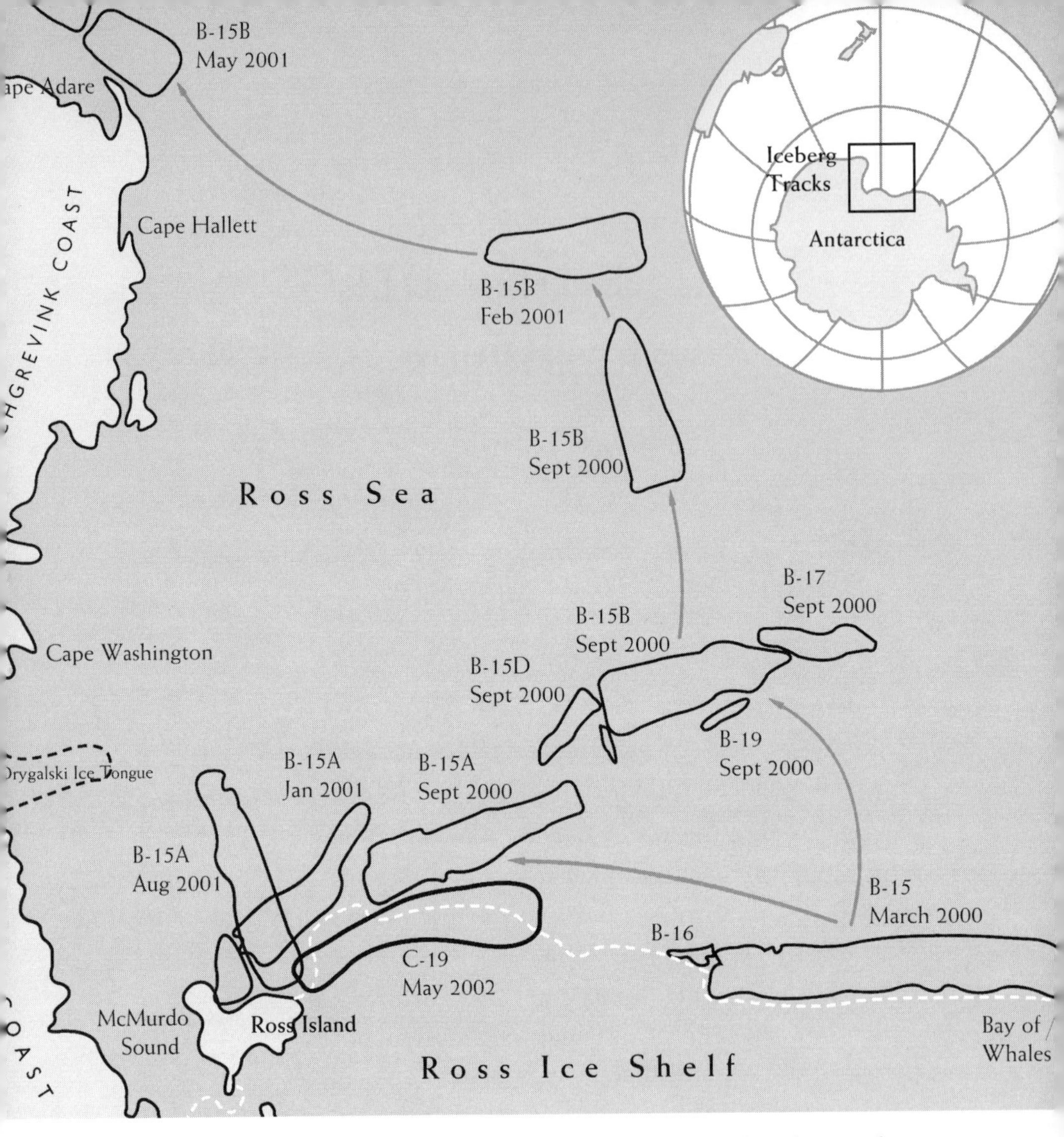

A series of satellite images transmitted to the Antarctic Meteorological Research Centre of the University of Wisconsin enable tracking of B15-series icebergs since the first breakout was detected in March 2000. Movement of the 'bergs around Ross Island means that many of the region's Emperor and Adelie penguin populations are denied access to their normal feeding waters. Usually about four-fifths of an iceberg lie below water level.

almost as large as the State of Connecticut, or a sixth of the size of Tasmania, or more than four times that of the Australian Capital Territory; depth was estimated to vary from 350 feet (100 m) at the seaward side to over 1,100 feet (350 m) towards the middle.

Scientists at the University of Wisconsin calculated that B15 contained 480 cubic miles (2,000 km^3) of ice, which, if melted, would produce 528 trillion gallons of water, sufficient to supply all of the United States' needs for five years.

Two years after B15, another huge iceberg broke in May 2002 from the Barrier's western edge. Classified as C19, it was 125 miles (200 km) long and 20 miles (32 km) wide.

Voyage to the End of the World

Adelie penguins cluster on a floe when Kapitan Khlebnikov *pauses at Cape Adare. The majestic peaks of the Admiralty Range stretch across the horizon behind the ship.*

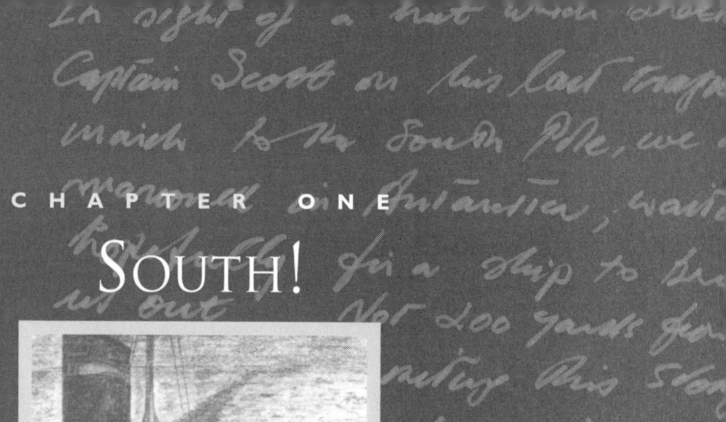

CHAPTER ONE

South!

Three days short of a new year dawning, with a full complement of ninety passengers from ten different nations, and with sixty officers and crew, the mighty icebreaker *Kapitan Khlebnikov* headed out from Hobart into the stormy Southern Ocean with a goal of sailing farthest south into Antarctica that a ship had ever reached.

Five days of constant rolling and pitching of the 12,000-ton ship (icebreakers are round-bottomed and have no keel!) lie in ambush through the twenty-five-foot (eight-meter) waves of the roaring forties, furious fifties, and screaming sixties; fifty-knot winds and up to forty-five-degree rolls are among the fun memories—little wonder passengers and crew enthusiastically welcomed the sight of the first drifting iceberg, a promise that the pack ice and consequently calmer seas lie ahead.

71°15.7'S:

The course takes *Khlebnikov* across the Antarctic Circle and past mist-shrouded Balleny Islands before reaching the coast of Victoria Land at Cape Adare, where in 1898 the Swedish-born explorer Carsten Borchgrevink and his Australian expedition were the first to winter-over on the Antarctic continent. In calm sunny weather, Borchgrevink's restored hut and its thousands of Adelie penguin neighbours are only a brief Zodiac ride away. In the background rises the majestic Admiralty Range and beyond lies the border of Australian Antarctica.

72°17.2'S:

Pointing its spoon-shaped bow ever southwards, the big ship from Vladivostok breaks a "garage" in the fast ice at Cape Hallett, allowing passengers to walk down the gangway and, braving the rising snow flurries, follow a line of orange flags towards the blue-tinged caverns of a grounded tabular iceberg.

74°39.4'S:

Then onwards again, manouvering through thick pack towards Cape Washington, pausing to inspect the few remaining inhabitants of an Emperor penguin colony that lies about a half hour's trudging across the frozen sea; the visitors return for afternoon tea served on the ice, and to report that for a wistful group of orphaned Emperor chicks, the outlook isn't so good.

Left: Douglas Mawson, who lived at Cape Royds as a member of Shackleton's 1907–1909 Expedition. Right: The Swedish-born explorer, Carsten Borchgrevink, leader of the 1898–1900 Australian Expedition that was first to winter-over on the Antarctic continent.

Left: Sir Ernest Shackleton—"the boss." He led the British Antarctic Expedition to Cape Royds.

75°23.0'S:

Among the peaks of Victoria Land, the lofty dome of Mt. Melbourne, a dormant volcano, plays hide-and-seek through the clouds as we cruise along the endless front of the Drygalski Ice Tongue, a glacial outpouring so large, we are told, that it can be discerned from outer space. Greetings are exchanged with the welcoming committee of Antarctic inhabitants: crabeater, leopard and Weddell seals; minke and killer whales; Adelie and Emperor penguins; and a growing armada of polar birds. Nine days away from Hobart and through the thickening Ross Sea pack we are closing on McMurdo Sound, entering the coast of the heroic age of Antarctic exploration. *Khlebnikov* with all 24,000 horsepower at the captain's bidding cuts for itself another dock in the fast ice; the gangway is lowered, the blue-striped helicopters made ready, our guides beflag a safety path and the tread of heavy boots begin to leave their mark on Ross Island snows.

77°36.7'S:

Around the outcrop of Cape Royds and above the rocky slope of Back Door Bay is found Antarctica's most southerly Adelie penguin rookery, alive with their scrambling, squabbling, squawking, and smells. Just beyond the rookery, rather like a stranger trying to avoid attention, stands the modest one-room hut of arguably the greatest leader the Antarctic has known, Sir Ernest Shackleton. We are on the very slope walked by men of the British Antarctic Expedition of 1907–1909. From here a young Douglas Mawson, with his mentor at Sydney University,

Above: Adelie penguins find a useful lodging among the remains of old Australian stores at Cape Adare. Right: The view from Ridley Beach towards Borchgrevink's preserved hut and its penguin neighbours.

Professor Edgeworth David, went out to climb the volcano Mount Erebus, which almost casts its shadow on the hut at Back Door Bay. From here, Mawson, Professor David (then a man in his fifties), and Dr. Alistair Mackay began their march to locate the South Magnetic Pole. And perhaps most memorable of all, from here Shackleton and his three-man team made a bid for the Pole and failed

The hut above Back Door Bay, headquarters of the 1907–1909 British Antarctic Expedition. New Zealand authorities care for the historic Ross Sea sites that lie within the Ross Dependency.

to reach it by just 97 miles (155 km). For "the boss," turning back must have been a decision that called for as much guts as the very trying to attain the Pole, but rations were dangerously low and turn back he did in the cause of survival. We wonder how often Shackleton pondered that a camp on the Bay of Whales, 60 miles (96 km) closer to the South Pole, might have made all the difference? A helicopter ride, courtesy of Vladi-vostok Air, to Cape Evans 5 miles (8 km) farther down the Sound leads to the much larger hut of the British Antarctic Expedition of 1910–1913 and the haunting memory of a polar tragedy. The long table where Herbert Ponting photographed Captain Scott sitting with his men to celebrate a mid-winter's feast is just as they left it, and so is everything else: boxes of groceries, tins of matches, candles, clothing, even the still-oozing slabs of Weddell seal blubber—all carefully restored in recent years by the New Zealand Heritage Trust. Rob, one of our number, is the official New Zealand government emissary who carries the key and, ever watchful of our behavior, unlocks all the huts and invites us inside. Antarctic Treaty notwithstanding, the New Zealand government is not about to let us forget that we stand on their Ross Dependency. "The hut has its ghosts," says Bob, our lecturer from Cambridge University, and is quick to add, "but they're friendly ghosts." Through this narrow doorway, Cherry-Garrard with "Uncle Bill" Wilson and "Birdie" Bowers in the depths of winter began the worst journey in the world—to steal a couple of eggs from the Emperor penguin rookery at Cape Crozier. From here on a windy yet more salubrious November day, Scott led his men to plant the British flag at the South Pole. They reached the Pole on 17 January 1912, only to find Amundsen's tent and letter awaiting them. The Norwegians, moving expertly on skis and with dogs by the dozen to haul their sledges, had won by a month.

Scott, with Wilson, Bowers, Oates ("I am just going outside and may be some time"), and Seaman Evans, turned wearily for the 800-mile (1,280 km) slog back to McMurdo. They were man-hauling, exhausted and dispirited. "There is that curious damp cold feeling in the air," wrote Scott, "that chills one to the bone." The weather turned to a death trap and never let them go.

At Cape Royds, Shackleton's dog kennels still survive the blizzards.

77°51.1'S:

Through the final 30 miles (48 km) of McMurdo's channel, and with all six diesel-electrics beating out maximum horsepower, we continue as far as the ice-filled waters will allow our double-thickness hull to push in this far corner of the Ross Sea. Winter Quarters Bay, where Scott landed on his first expedition in 1902, marks the end. The Discovery hut, named after the expedition vessel that was iced-in here two winters through, nestles above the Bay. As with the others, the hut has been carefully restored, a far cry from 1958 when I was on my first visit at McMurdo and it lay derelict yet mercifully semi-embalmed within a block of ice. (We pause for a humorous aside—back in 1911 when Scott found the hut blocked with snow, he blamed Shackleton for leaving a window ajar in 1908.) Planned more as a storehouse than for occupation, the Discovery hut is a shadowy place where one should listen to the whispers. Mawson and others who were with Shackleton in '08 rested within. I have a letter from Sir Douglas Mawson that reads, "I well recall camping there for a few days' rest at a time during my year as a member of the first Shackleton Expedition." Scott, of course, used his hut in 1911; it would have been his last shelter, offering safety and protection before marching south across the Ross Ice Shelf. Shackleton's rather forgotten Ross Sea Shore Party camped here in 1914–1917, wondering if "the boss" and his Trans-Antarctic Expedition would ever come, unknowing that an incredible drama of a lost ship, survival, and rescue was being enacted on the far side of the continent. At the cost of one life, they had laboured to lay supply depots as far as Mt. Hope at 84°S. Two others, Mackintosh and Hayward, dared the sea ice on 8 May 1916 to reach Cape Evans, never to be seen again. Theirs is the memorial cross above the Cape. Not content with one enormous adventure, Shackleton himself returned to retrieve the survivors.

Where *Discovery* once anchored, the U.S. Coast Guard 'breaker *Polar Sea* is moored at the ice wharf, awaiting a call to clear the channel for thin-skinned ships bringing fuel, heavy equipment, and provisions to McMurdo, some to be flown onwards to the Pole. We helicoptered across the choking ice to new McMurdo Station, which through fog greets our vision with rows of drab brown three-story dormitories,

Above: At Cape Evans, Scott's men celebrate their 1911 midwinter dinner. Scott sits at the head of the table; he and three others at the dinner (Wilson, Bowers and Oates) would not survive the South Pole march. Right: The same table as it is kept today in the expedition's preserved hut.

networks of pipes, power lines, parking lots packed with trucks and tractors, fuel tank emplacements, and scurrying rugged-up men and women. For a veteran who first lived here over forty years ago in a rough settlement of canvas Jamesways and metal Quonset huts, stepping from the 'chopper is quite a shock to the system. Forget that the South Pole is only a few hours' flying time away by ski-equipped C-130 Hercules; today's McMurdo has all the architectural charm of an outback mining town. But through summer, the local population expands to over 1,000 people, so what should you expect?

This afternoon we are off to meet the Kiwis. The helicopter ride leads by Observation Hill from where they searched the dazzling white of the Ross Ice Shelf for a trace of marching men, only to report back at the hut "the Pole party has not returned." The tall jarrah cross in memory of those who perished is clearly visible from the 'chopper window. Climbing the hill (as I once did in gathering darkness) leads one to read the names of Scott and his companions, and Cherry-Garrard's choice of lines from a verse in Tennyson's "Ulysses"—"to strive, to seek, to find, and not to yield." Modern Scott Base is a series of bright green buildings; neat and inviting they replace the original "plywood" huts where Fuchs and Hillary climbed out of their Sno-Cats at the end of that "last great journey open to man," the Trans-Antarctic Expedition of 1957–1958. But if Scott Base gives a snuggle-up feeling, one should remember that the summer population is about 50 and in winter reduces to around 10; at McMurdo Station the Americans have 200 wintering-over.

In bright evening light—night counts for nothing at this time of the year along 77 degrees South—the ship cuts a path westwards across the Sound towards the magnificent rim of the Royal Society Range. Once more, to commuter-like regularity, the helicopters are readied on *Khlebnikov*'s rear deck to carry intrepid travellers across the ice to confront an Antarctic contradiction.

77°30.0'S:

In twenty minutes they are landing amid an eerie landscape where feet tread unexpected rock and hardened soil. This is the Taylor Dry Valley, stretching westwards towards the Plateau, hemmed in by the snow-topped Kukri Hills and the Canada and Commonwealth glaciers, their outfalls halted in icy immobility; the remains of a mummified seal add to the mystery. Griffith Taylor, though born in England, was one of the two young Australian scientists who joined Scott's second expedition. Frank Debenham, born in an Australian country town, was the other. Both complained of sometimes being treated as "colonials." Debenham, who later became a professor at Cambridge, wrote in evident distress to his mother, telling her of a fallen idol: "I am afraid I am very disappointed in him, tho' my faith

Catherine brings yoga to the Taylor Dry Valley. Beyond the author's wife lie the immobile cliffs of Canada Glacier.

has died very hard . . . His temper is very uncertain and leads him to absurd lengths even in simple arguments. In crises he acts very peculiarly . . . What he decides is often enough the right thing I expect, but he loses all control of his tongue and makes us all feel wild . . . I cannot say he is in the least popular." Such was his reluctant judgment of Robert Falcon Scott.

Above: A Dry Valley mystery—remains of a seal resting far from the sea.

Right: In fine weather, Khlebnikov's *bows offer travellers a choice vantage place to watch Antarctica and its ice-choked seas unfold before them.*

77°37.7'S:

The hour is late when everyone is safely back aboard, reflective on the riddle of the Dry Valley's moon-like landscape, wondering about a stranded, long-dead seal and giving thanks for the trio of Austrian chefs who create daily menus that would do proud to many a five-star hotel. *Khlebnikov* is meanwhile making a 180-degree turn and, at last, heading towards one of the world's great natural wonders, the 450 miles (720 km) of the Great Ice Barrier. At the far end of the towering white cliffs, our goal of Farthest South at the Bay of Whales is waiting. And soon we will have an appointment with B15, one of the largest—if not the largest—Antarctic icebergs that has ever been

observed. The climax of a voyage to the end of the world comes closer.

My first sighting of the Bay of Whales had been back in 1964, from the cockpit of a U.S. Navy aircraft after an incident-prone pioneering flight ("prepare for crash landing") from Melbourne, Australia, to the geographic South Pole. Beneath our C-130 Hercules lay the huge white hump of Roosevelt Island against which ice flowing from the plateau across the Ross Shelf is forced to divide, and so forms a Bay whose only claim to permanence lies in its ever-changing, ever-unpredictable shape. But eight years before then I had made my first icebreaker voyage from Antarctica and across the Great Southern Ocean—a wild experience, never to be forgotten, as I recorded in a feature article for the *Sydney Morning Herald*.

> The captain stumbles across the wardroom, throws his arms around the neck of the junior lieutenant, mutters an apology and crashes back against the wall. The chief engineer drops into a chair, fastens a safety belt around his middle, presses his feet against the floor and wards off a sauce bottle that comes slopping towards him. The gunnery officer, his meal over, makes a dash for the door—wondering whether he'll end up in a bruised heap at the far end of the companionway, or flop into the cup of coffee the padre's balancing on a knee.
>
> This is life aboard the *USS Glacier*, the world's roughest, toughest ship. *Glacier* is the biggest ice breaker in the "free world"—8,600 tons of rolling, rocking metal, her hull two inches thick, her crew some of the gamest sailors on the seven seas. South from New Zealand her beat is across wild, grey, lonely oceans—the roaring forties, furious fifties, and screeching sixties—on into the deathly seventies, where icebergs the size of a city drift off a frozen continent. Each year she is the first ship into Antarctica, and the last to get out, and the blue

channel she carves through the ice is the highway along which the men and materials of Operation Deep Freeze ride. She rocks so much that her passengers almost walk up the walls. Her crew sleep with their bunks chocked up in a "vee"; all her chairs have safety belts to fasten them to the tables; and to fall overboard from her icy decks means death in four minutes in the freezing Antarctic waters—if the killer whales don't get you first.

I came back from Antarctica aboard her, on an unforgettable 2,200-mile last voyage out of McMurdo Sound that still lives in my memory like a lurid dream.

First sight we had of *Glacier* was a ghostly ship, wreathed in "sea smoke" rising off the freezing waters of the Sound. ["Sea smoke" is an Antarctic phenomenon caused by waters being warmer than surrounding air.] The temperature was down to minus 56 (Fahrenheit) and we were marooned. All flying for the season had stopped, the ice runway had crumbled, and here on the Sound—the nearest point that a ship can sail to the South Pole, lying only 799 miles away across Barrier and mountains—*Glacier* had broken in to get us out.

For two days she hove-to off shore, battling against a shrieking 63-knot blizzard. Her two scarlet helicopters were literally frozen to the deck, and as fast as they steam-hosed her steel landing barges free of their davits, condensation from the steam turned back into ice! The barge that finally made it was crewed by men with masks on their faces. Two hours is the maximum they can work outside in the weather. We climbed slippery, treacherous jacob's ladders to a deck thick with the crunchy ice that coated her plates and hung in white stalactites from the rails and rigging, some of it spearing down on our heads. AGB4 was the designation written in black letters on her bow plates. To the crew, that means the "Great Buster."

But below decks, she was unbelievably warm and cozy . . . our first real shower in weeks, first real cup of creamy coffee, first change of clothing, first proper bed with sponge rubber mattress.

"A luxury hotel," we thought—fools!

Glacier's captain, "Sam" (Joseph) Houston, called a conference in the wardroom. "We're going to break into Cape Hallett tomorrow night," he said. "It'll be risky, because it's late in the season and there's a helluva lot of ice about. We'll go in and get out just as fast as we can." Dark eyed, with a wide, handsome grin, our Bostonian skipper is vice-president of the self-styled "Ross Sea Ferry Boat Company." He might well be Hollywood's idea of what an icebreaker captain should look like, but more importantly he seems to be a captain who knows his job.

First warning we have of the ice that bars our escape are the "lilypods," then the pancake ice, then the bergy bits that come floating past, like fugitives from a toyland nightmare. Icebergs are on the horizon. Sinister and silent, they are flat topped, a mile long maybe, a hundred miles maybe, drifting like deserted aircraft carriers across this pitiless sea. In their jagged sides they have shadowy caverns, large enough to swallow our ship whole.

"Ice ahead!" The cry goes up from the bridge, where radar and sonar continually probe with invisible fingers. It is the pack, glittering white, stretching ahead far as the eye can see. Like the icing on some gargantuan cake, it is twelve feet, maybe fifteen feet thick, the sky above it filled with the ice blink, the telltale reflection that was a warning beacon to the mariners of old.

"Full ahead," calls skipper Houston. His junior lieutenant slams over the engine telegraph. In a matter of seconds, *Glacier* is throbbing ahead under the full thrust of her 21,000 h.p. engines. Ice like this can crush a ship as though it were tinplate. It can beset a ship and lock it in so that they'll only find the skeletons. We are going to punch a hole right through it.

Glacier charges in like a bulldozer at full throttle, her great spoon-shaped bow rising up, the whole ship filled with the vibrations of her 15 diesel-electric motors. Seals, wide- and stupid-eyed, flop away from almost beneath the prow. Adelie penguins flee in panic across the floes, waving their little flippers.

Suddenly it happens.

"Blump-blump-blump" goes the ship as steel rams ice; the shock of it is like being in a truck jolting across an outback road. Ugly black scars criss-cross the whiteness. It disintegrates in searing, hissing chunks under the impact of our hull, turns over with the gurgling rush of boiling waters, exposing beneath it a brown coating of life in these rich Antarctic seas.

"Full astern!" calls Houston, and *Glacier* goes fast into reverse.

"Full ahead!" She charges the pack again.

"Full astern!" So it goes on all day and night as the ship tears out an enormous wedge-shaped channel down which she can safely sail without the ice freezing in behind her. One 'berg alone we encountered had an estimated weight of 5,000,000 tons . . .

In the dead of night *Glacier* creeps into Cape Hallett, last stop on the Antarctic mainland before the coastline fades away on our port-side to the frozen perimeter that curls away for thousands of miles, far to the south of Australia.

"The most scenic spot in the Antarctic—the Banana Belt of the Pole," they call this postage-stamp fifteen-man base perched on an ice ledge with the frozen cliffs of Victoria Land rising sheer behind it. Twin searchlights from our bridge plunge into the night, catching a bombardment of snowflakes in their beams. The wind offshore cuts like a knife. It is so cold that we feel the words almost freeze on our lips.

This is a difficult enough place to reach at any time, and why have we come in when the Antarctic sailing season should be well over? Supplies? Science? Evacuation? No, just toothache. Two men at the base have lost their fillings, and to leave them without treatment through an Antarctic winter would be akin to the cruelty of the medieval thumbscrew and the rack.

Beyond Cape Hallett, in the open seas that swirl unchecked around the bottom of the world, *Glacier* is a mad thing, rolling and heeling relentlessly, until her dizzy-headed passengers wonder if the world were ever still. Every ten seconds she rocks; side-to-side, side-to-side, sometimes to forty-five degrees. One moment a porthole mirrors the blue sky; next moment it is an eyeglass pressed deep against the white-flecked waves that go foaming past, not an arm's length outside. Falling crockery, slithering chairs, creaking metal are the noises of her progress as she forges through the waves at a steady seventeen knots. The man you are talking to is suddenly catapulted out the door and last you see he's disappearing through a hatchway; officers sitting on opposite sides of the wardroom hurtle together in the middle of the floor, still sitting in their chairs.

Will it ever stop, you wonder?

Frank Debenham, the first Australian-born scientist to go to Antarctica, described a view of the Great Ice Barrier from the deck of Scott's second vessel, *Terra Nova,* in February 1911:

> The pack ice extends several hundred miles southward from Scott Island. Ships break through it into iceberg-filled open water of the Ross Sea, a part of the ring, which nearly encircles the continent. Before them looms a wall of ice, from fifty to a hundred feet high and

The ice-coated decks of USS Glacier *warn that winter is fast approaching on the Ross Sea.*

five hundred miles long; it is unscalable and impassable. The glittering white cliffs, ramparts looming over the cold waters, provide one of this planet's most awesome spectacles.

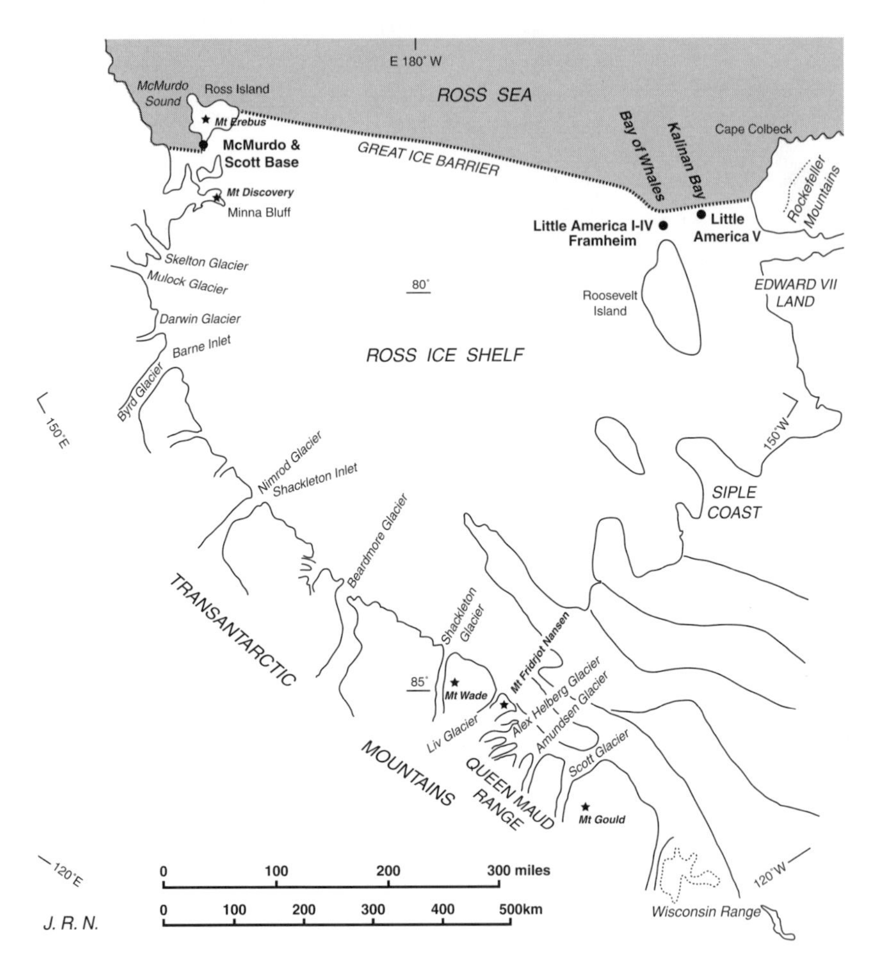

And of the gargantuan forces that give birth to the Bay of Whales, Debenham went on to explain:

> Once around Roosevelt Island [a rock hump of about 1,000 feet elevation that rises in the ice shelf behind the Bay of Whales, dividing the glacial flow] the two shelves tend to coalesce again. One mass of about sixty trillion tons and another of twenty trillion tons move inexorably toward each other across the narrow bay of open

Masked against the cold, a Navy crewman from USS Glacier awaits McMurdo's last outbound passengers.

water filled with floating icecakes at a speed of three or four feet a day. Every few decades the two masses crash, with about the most stupendous manifestation of force on earth. A great chunk is cracked from one or the other—which mass suffers depends on internal stresses at the time—and floats northward as an iceberg, or several icebergs. When this happens, the temporarily closed Bay of Whales is recreated, with the size of the indentation at any given time depending on the size of the chunk and the time that has elapsed since the crash.

Debenham's "stupendous manifestation of force" has happened. In its aftermath, aboard *Kapitan Khlebnikov* we follow a course impossible for men of the heroic age and the later-twentieth-century explorers. Even the redoubtable *Glacier* couldn't have cut its way here. The calving of B15 has created a new, more southerly Barrier front, yet to be mapped. We are possibly sailing an icy sea where no ship has ever been.

Heavy pack off Cape Hallett calls for the dispatch of Khlebnikov's *helicopters to scout for open water leads, known as* polynyas.

CHAPTER **TWO**

Amundsen's Enemy

At Hobart, Tasmania, on Thursday, 12 March 1912, a scruffily-dressed figure rowed ashore from a strange vessel that had anchored in the fairway of the Derwent estuary. Avoiding watchers at the quay, the figure booked into Hadley's Orient Hotel. "Treated as a tramp . . ." he afterwards noted in his diary, "in my peaked cap and blue sweater—given a miserable little room."

Within the hour the same figure was at the General Post Office, sending a coded cable to his brother in Christiania. Thus did the world learn from Roald Amundsen about Norway and the Pole.[1]

Roald Amundsen, a hero of Arctic exploration, did not share Shackleton's fear that he would dig his grave at the Bay of Whales.[2] Close studies of the diaries of Ross, Borchgrevink, Scott and Shackleton gave the Norwegian adventurer, who had already once wintered in Antarctica, the confidence he needed to steer *Fram* through the gap in the Great Ice Barrier.

Above: An Australian stamp honours Fram, *the famous Norwegian diesel-powered vessel that brought Amundsen to Antarctica. Right: Roald Amundsen.*

A huge and mostly ice-free embayment awaited his ship; better, at the far southeastern end of the Bay a small inlet offered a secure anchorage—and Amundsen knew that every mile he saved in the march to the Pole could prove vital. On 15 January 1911, *Fram* moored at the edge of the bay ice and in weather described as "brilliant, sunny, and warm," preparations were begun for the task of unloading. In Amundsen's words:

> Here on the same barrier where Shackleton praised his God that he had not landed—here we will have our home. That Ross did not want to come too close to this ice giant in his sailing ships—that I understand. But that S. did not come here and take the great chance offered by an extra degree of Southern latitude; that I don't understand. Not one of us has given a thought to any danger in doing so. The future will show if we were right.

About four miles from the ship they found a relatively sheltered position at the crest of the ice shelf where Framheim, the dismantled hut built in Norway, could be erected. Their ninety-seven Greenland dogs,

eager for exercise after four months aboard ship, bounded up the snow slope with the laden sledges in tow, and in three weeks ten tons of supplies and equipment were in place. In that time, they reckoned, each of

Erebus

Captain James Clark Ross RN first sighted and named the Great Ice Barrier on 28 January 1841; the record from his epoch-making voyage reads:

"As we approached the land under all studding sails, we perceived a low white line extending . . . as far as the eye could discern eastward. It presented an extraordinary experience, gradually increasing in height as we got nearer to it, and proving at length to be a perpendicular cliff of ice. Between one hundred and fifty and two hundred feet above the level of the sea, perfectly flat and level on top and without any fissures or promontories, even on its seaward face. We might with equal chance of success try to sail through the Cliffs of Dover, as to penetrate . . . the icy barrier."[1]

Sailing the Barrier and entering the (as yet unnamed) Bay of Whales made a dramatic impact on the hardened seafarers of Ross's company in *Erebus* and *Terror:*

"I kept the deck throughout the night, a night never to be effaced from memory's tablet to the latest hour of existence," wrote the surgeon of *Erebus*, Robert McCormick. "At 5:40 A.M. the break in the barrier forming an inlet or bight, perhaps, a quarter of a mile wide, and from a mile or two miles in depth, bounded on its starboard side by a very strikingly bold promontory of ice, for which we had been for some time standing in, and now went about when within a quarter of a mile of it, with a moderate breeze blowing.

We had tacked none too soon for its great height above our mastheads, even at this distance, took the wind out of our sails as we hung in stays."

Cornelius Sullivan, the *Erebus* blacksmith, recognized in the

Erebus and Terror locked in Antarctic ice during Captain James Clark Ross's voyage of discovery, 1840–1843.

Barrier "nature's handiwork" as he described at 78°04'S their approach to the wall of ice:

"We pursued a South Easterly Course for the distance of three hundred miles. But the Barrier appeared the Same as when we Left the Land. On the first day of Febry. we stood away from the Barrier for five or six days and came up to it again farther East, on the morning of the eight Do. We found ourselves Enclosed in a beautiful bay of the barrier.

"All hands when they came on Deck to view this the most rare and magnificent Sight that Ever the human eye witnessed Since the world was created actually Stood Motionless for Several Seconds before he Could Speak to the man next to him. Beholding with Silent Surprise the great and wonderful Works of nature . . . I wished I was an artist or a draughtsman instead of a Blacksmith and Armourer."

the dog teams would have covered 500 miles (800 km) making the journeys to and fro.

Olav Bjaaland, aged thirty-eight, was Norway's champion skier; from their mooring he gazed at the immensity of the Barrier to record this impression:

> It is a strange feeling that grips one as the sight now reveals itself. The Sea is still as a pond, and before one stands this Great Wall of China and glitters. Far off, it is like a photograph that has just been developed on the plate. By letting one's thoughts wander over the surface, one finds oneself in a melancholy mood. One thinks of what is to come, the hardships one is going to meet, the use one will be, and if we can get there before the Englishmen—who are surely burning with the same ambition.[3]

On 15 February they farewelled *Fram*, watching Nansen's famous old ship make a circuit of the clear innermost waters of the Bay, where it reached 78°41'S, a southerly sailing record that has not been bettered. After a winter spent in Buenos Aires, the ship was scheduled to return in a year to take them home.

Amundsen chose eight men to remain with him at Framheim, and among them were the three, four or five who would be selected for the great push south. Each of them would be an expert at skiing, navigating, dog handling, and possibly all three; they comprised, it could be said, the pick of Norway's polar travellers. Before darkness

Framheim, the Norwegian base of 1910–1912 at the Bay of Whales.

Above: Strange encounter at the Bay of Whales on 18 January 1911. Aboard Scott's Terra Nova *(left) geologist Raymond Priestly recorded "astounding news" of another ship, recognized as* Fram, *anchored beside the Barrier.*

descended they set out with the dog teams to lay supply depots at every degree until 82°S; in three weeks, one and a half tons of provisions were moved, the last being within 480 miles (770 km) of the Pole. Then with ice forming across the Bay, the huskies secure on the dog lines and a long list of equipment to be readied, they waited to see the winter pass.

Amundsen's very coming to Antarctica was a drama in itself. He sailed with nineteen men aboard *Fram* ostensibly to conquer the North Pole. But the world learned that on 6 April 1909 Commander Peary, the American, had reached the Pole; at this news a thwarted Amundsen did not scrap his plan, he simply changed it.[4] Only to his captain, Lieutenant Thorvald Nilsen, did he first confide there was no point

Left: Dog teams were the key to the Norwegian's rapid passage to the South Pole. Helmer Hanssen with his team of Greenland huskies at the Pole, 15 December 1911.

in undertaking a voyage that had been milked of its challenge. At Madeira he told the ship's company that it was not the North but the South Pole they were seeking. The explorer's magnetic personality was in full play. Every man accepted the decision, even though it meant long months at sea and a year or more in the remotest corner of the far end of the earth. Respect is what Amundsen received from his men. Like a true Norseman in spirit as well as profile, he was not given to a show of emotion. Aged thirty-nine, the conqueror of the northwest passage stood only one step beneath Fridtjof Nansen, who was hailed as the greatest of all Norwegian explorers. At Framheim, where he answered to the "The Chief," the mood was informal, even-handed and in general unmarked by controversy—except for the influence of one figure, Hjalmar Johansen.[5]

Johansen had been Nansen's companion on the northern expedition and acquitted himself as tough and able amid the drifting Arctic ice. To have served alongside the national hero—indeed, at one moment he had saved Nansen's life—fuelled Johansen's belief in himself and a doubt that any other leader, Amundsen included, could prove his superior. But fame at Johansen's level had a limited endurance and the disappointed champion skier and dog handler lapsed into heavy drinking, which in turn destroyed his marriage, his army career, and his savings.

Nansen, from whom Johansen had been borrowing money, wanted to remove his old comrade from the temptations of life in Norway.[6] At his request, and against his own better judgment, Amundsen included this undoubted "rival" in the team at Framheim. But in isolation and winter darkness the potential friction between the two men was impossible to hide. Johansen was a short muscular man whose very presence and past relationship with Nansen embodied a threat. Amundsen found him questioning, critical and sometimes non-cooperative; the thought that Johansen might undermine his leadership to the extent that he faced a mutiny in the expedition was a real fear in Amundsen's mind.

A particular stormy scene followed Amundsen's risky and failed attempt to start for the Pole on 8 September when temperature on the ice shelf fell to a punishing –50°. Johansen claimed he had saved at least one of Amundsen's men from being lost in the disorganized dash back to Framheim and next day, with some justification, gave vent to his disgust. Amundsen's diary records the outburst:

> At the breakfast table . . . Johansen found it suitable to utter unflattering statements about me and my position as leader for our enterprise here. It was not only our return yesterday that he found indefensible in the highest degree, but also much else I had taken the liberty of doing as leader in the course of time. The gross and unforgivable part [of Johansen's statements] is that they were made in everybody's hearing. The bull must be taken by the horns; I must make an example immediately.

Next day Amundsen announced that Johansen was excluded from the Pole party; instead he would join a subordinate expedition sent towards King Edward VII Land. The dream of once again enjoying the fame and possibly the fortune of participation in a great exploring feat vanished. Johansen had been dealt a blow to his self-esteem, one from which in truth it seems he never recovered. A year later he shot himself in a little room in a cheap Christiania hotel; the only belongings found with his body were an empty cigar box and his shaving gear.

Some were moved to ask, perhaps unkindly of Amundsen, how much the episode at the Bay of Whales had contributed to Johansen's suicide. Certainly there is reason to believe that this death and those of Scott and his companions in striving for the Pole were scars that remained on Amundsen's memory across the years ahead.

For him, fame came with bitter dregs. He returned to find that the woman he loved would not divorce her husband, and a bachelor he remained until he was last seen disappearing into the Arctic skies.[7] The British gave him scant respect, they called for "three cheers for the dogs" and in some schools still taught that Captain Scott discovered the South Pole. He quarreled with his brother Leon, who was also his manager over his finances, which were forever in a mess. And no other great expedition would await him to equal Antarctica.

Personalities were forgotten on 20 October 1911 when Amundsen with his expert team of Bjaaland, Oscar Wisting, Sverre Hassel, and Helmer Hanssen, all on skis and with fifty-two yelping dogs dragging the four laden sledges, left for the south. Amundsen made no pretence that his was a scientific expedition. They were explorers and, true to the dedication of their legendary ancestors, the Pole was the prize. Their course made an almost direct line from the Bay to the Pole; crossing the Ross Ice Shelf, ascending the glacier, and driving across the plateau with almost surgical precision.

With the wind frequently behind them they were able to make 20 miles (32 km) or more a day, sometimes riding on the sledges and letting the dog teams have their head. The grim reward for this unbounded canine enthusiasm was a bullet and a chopping block until with the half-ton loads lightened, only eighteen dogs remained to see them home. En route they shared in the perils of Antarctic travel, straying into a crevasse belt being among them. "I was riding with Wisting," Amundsen wrote. "His sledge was the last. Suddenly a large piece of the surface disappeared, and a gaping abyss opened next to the sledge—big enough to swallow us all." Finding a way up the tortuous Axel Heiberg Glacier tested all their skills. On the polar plateau, in the face of a razor-like wind, frostbite began to show. But

COLMAN'S
CORN-FLOUR
No.1744
B.A.E.
SHORE
PARTY
COLMAN'S
CORN-FLOUR

Left: Provisioning Scott's hut at Cape Evans in January 1911. Above: The same hut as it appears today, under the protection of the New Zealand Antarctic Heritage Trust.

these were incidents along the way—they escaped the full fury of the Antarctic weather that was reserved for those who followed a month later, who could only exclaim "Great God, this is an awful place!"[8]

On 26 January 1912, heavy footsteps intruded at 4 A.M. upon the slumbering Framheim. Lindblom, the cook and watchman, opened his eyes to find a smiling Amundsen gazing down at him. More tramping sounds as others of the party came thumping in, shouting for hot coffee. At last someone in the hut dared ask, "Did—did you make it?" The answer was a resounding "yes!" On 15 December they had unfurled the flag at the South Pole. Of the British they found no trace. Norway had won the Pole, *Fram* was back in the bay, and now they could all go home.

When the Americans established their South Pole Station in 1956–1957, they named it Amundsen-Scott in honour of the two leaders who first reached the Pole, one of them never to return. Amundsen took all of a few days short of two months to attain his goal and for Scott it was an even longer march; forty-six years later air transport had reduced the time to a matter of hours, as I wrote for the *Herald* of a supply drop to the bottom of the world.

> "We're coming over the Pole! Open doors!" the Globemaster captain raps into his microphone.
>
> "Doors open. Parachutes ready," the voice of the drop master crackles loudly back on the headphones.

"Approaching target zone . . . Stand by . . ."

"Standing by . . ."

Four engines trailing vapour, the plane roars above the circle marked out with fuel drums and flags on the glittering ice. Shielding their eyes, the little group of men in snow hoods and parkas look up at the silver and scarlet-tipped hull and wings.

"Twenty seconds, 15 seconds, 10, 5, 4, 3, 2, 1."

The captain's voice rises.

Then: "Zero!—Drop!"

"Drop."

Inside the fat metal belly the drop master presses a button. "Zing! Zing! Zing!" go the parachutes as they skid down the fuselage on wire runners overhead: a screeching wind plucks them through the hole: the static line catches and they billow out in huge umbrellas of vivid red, yellow and green nylon.

"All away and on target."

"O.K. Close the doors. We're going home." Motors roaring, the big plane is climbing, banking sharply and heading north—the only direction it can go from the South Pole.

Riding a Globemaster to Antarctica is like sitting in an airborne tunnel the size of an underground railway station—a tunnel stacked to the roof with generators and oil drums, mailbags and potatoes—whose metal sides vibrate continuously to the thunder of the mighty propellers outside. A narrow seat, about as comfortable as a park bench, running down one side is the only place to sit. A pile of supply parachutes heaped on the floor, the only place to sleep. The wingspan is 178 feet, nearly 30 feet more than the height of many a tall building. The scarlet tail is 48 feet high; the four engines each twice as powerful as a diesel loco. A travelling crane runs the length of the fuselage overhead. An elevator is fitted into the two-level floor, and mechanical loading ramps tuck into the nose that opens wide so that heavy transport wagons can be driven inside.

"First with the Most" is the motto of the "Blackjacks," the men of the U.S Air Force's 52nd Troop Carrying Squadron, who fly the Globemasters. From each October to February, their base is Harewood Airport, Christchurch, New Zealand, the jumping-off point for Operation Deep Freeze. South from Christchurch, they head 2,400 miles to McMurdo Sound, eleven hours away through the night to a mere pinpoint on the edge of the Great Barrier.

For remoteness and loneliness the flight has few parallels. The Southern Ocean tosses cold and empty below; the jagged, crevassed icy coastline waits ahead. The only emergency landing field is at Cape Hallett, 2,000 miles south. The only navigational aid is the weather ship *USS Brough* on station 800 miles south of Invercargill, sending out weather and radar signals. The "milk run" is a jaunty name for a flight far beyond the point of no return which even experienced aviators make "on their nerves." Men speak of writing long, endearing letters to their wives and families the night the Globemaster leaves; some hope the flight will be cancelled; some try to lose themselves in a book; some try to sleep all the way, with the help of pills; some call on the aid of Christchurch's inexpensive whisky.

No pretty, smiling hostess awaits you at the top of the ladder. "Can you swim?" snapped the zippered-up crew member the night I pushed my bags aboard at Harewood and climbed up after them through the narrow hatch. "Here's your survival gear." He pointed to a "Mae West" rubber immersion suit that was supposed to cover you from head to toe. (Would I have time to put it on?)

"If we ditch, the warning bell will ring six times. You put them on immediately." His voice has a frightening finality. "A continuous ring sounds one minute before we hit the water. Go to the rear door. There'll be a rubber raft waiting there. Don't fall in the water. You'll last about four minutes in the Antarctic seas."

From the runway at McMurdo, 6,000 feet long and bulldozed out of the ice, the Globemasters take off again, destination South Pole. Loaded with supply parachutes, they heave their eighty-five-ton bulks into the air and head 800 miles across trackless wastes. Beneath them is the desolate white of the Ross Ice Shelf; then the colossal winding highway of the Beardmore Glacier up which they swing, with mountains near 15,000 feet high rearing smooth and glassy peaks above them, fortresses guarding the vast interior plateau on which the South Pole lies.

Scott and his companions foot-slogged for two months on this journey. The Globemaster does it in four hours. At 12,000 feet altitude it is unbelievably cold above the Pole; temperature spirals down inside the plane; travellers can be issued with sleeping bags—not to sleep in—but to "wear." Smaller planes, like the Navy's jet-assisted Neptune bombers land on skis, to relieve the eighteen men permanently stationed at the Pole—but not the Globemaster with its heavy tricycle balloon-tyred wheels.

Each "drop" holds up to twelve tons. Huts, generators, fuel dumps, and food—all of it found at Amundsen-Scott Pole Station—has been built up by the Globemasters on the end of a parachute! In the rare Antarctic atmosphere parachuting is a risky business, the drop master and his assistants are highly skilled men. A caterpillar tractor on a 'chute that failed to open buried itself in a thirty-foot crater in the ice. It is likely to stay down there forever. Wind caught a generator dropped at Byrd Station and dragged it like a toboggan for twenty miles across the frozen plateau.

"Beg leave to inform you am proceeding Antarctic. Amundsen." A cryptic cable handed to him in Melbourne was the sum total of Scott's knowledge that the Norwegians were also heading south. So what was Amundsen's game? Last heard he was supposed to have the North Pole in his sights. Was the seasoned explorer, soon to be dubbed the "perfidious Norwegian," set on the Antarctic Peninsula, or the Weddell Sea or (surely not!) squatting in McMurdo Sound?

On 18 January 1911, Scott had his answer. Lieutenant Victor Campbell had been sent to take

Below: Fram *reached Hobart in March 1912, where the five men who had conquered the South Pole posed for local photographer J. Beattie. From Left—Sverre Hassel, Oscar Wisting, Roald Amundsen, Olav Bjaaland, Helmer Hanssen.*

Stores have been neatly replaced in Scott's hut at Cape Evans.

Terra Nova to the far eastern end of the Ross Ice Shelf, where a landing might be found at King Edward VII Land for a second exploring party of the British Antarctic Expedition.

Thick ice at the limit of the journey defeated the ship's progress. As a result, Campbell turned westwards again in the hope of finding "Balloon Bight" as recorded in Scott's 1902 *Discovery* journal, or the bay described by Shackleton in his *Nimrod* voyage of 1908—and of which (as with numerous of Shackleton's claims) various members of the Scott camp were highly skeptical.

As they rounded a prominence in the Barrier cliffs, the opening to a wide harbour spread across their view. Silently they gazed at almost ice-free waters of a majestic white amphitheatre, until a cry from the lookout told them that across on the far shore another ship was anchored.

Raymond Priestly, the expedition geologist, had been with Shackleton and in undisguised loyalty to "the boss" recorded his satisfaction at seeing that everyone was now forced into "backing up the Shackleton expedition."[9] His report of the event continued:

> I turned in . . . feeling quite cheerful and believing that there would be a good chance of . . . finding a home on the Barrier here . . . our last hope of surveying King Edward's Land. However, Man proposes but God disposes and I was waked at one o'clock by Lillie (the biologist) with the astounding news that we had sighted a ship at anchor to the sea ice in the Bay. All was confusion on board for a few minutes, everybody rushing up on deck with cameras and clothes. It was no false alarm, there she was within a few hundred yards of us and what is more, those of us who had read Nansen's books recognized the *Fram*.

But the Britishers' surprise differed little from that of the sole Norwegian watchman aboard *Fram* who, at the sound of "noises off" shortly after midnight, and fearing the Barrier was about to calve, raced on deck prepared for an emergency. Though quite relieved at the sight of *Terra Nova*, the watchman (as second mate, Lieutenant Frederick Gjersten described) preferred to take no chances with Englishmen:

> Our watchman . . . saw two men go ashore, put on skis and with reasonably good speed for foreigners, rush off up towards the barrier, following the dogs' tracks. "Well," thought the watchman, "if they have any nefarious intentions, (one of our constant subjects of discussion was how the Englishmen would take our challenge) the dogs will manage that job, and get them to turn round all right. It'll be worse if they sneak up to *Fram*, where I'm alone on watch. Best to be ready for all eventualities . . . " He dashed into the charthouse, carefully loaded nine bullets into our old Farman gun . . . dug out an old English grammar, and looked up "How are you this morning" and similar expressions. Thus armed to the teeth, both physically and mentally, he crept out on watch again. He might have been waiting half an hour . . . when suddenly a shock went through him. The

Heading down the Beardmore Glacier, a U.S. Air Force C-124 Globemaster returns from yet another twelve-ton parachute drop to supply Amundsen-Scott Station at the South Pole. The Pole Station, first occupied in 1957–1958, received 750 tons of equipment and provisions in supply drops from the big aircraft which made three six-hour return trips daily during summer season. Later, C-130 Hercules transports replaced the Globemasters, landing on skis at the Pole.

2990

> Englishmen were coming down again, with a course straight for *Fram* . . . He looked carefully: No, they had no weapons; if so, it must be a revolver in a pocket . . . He put down the gun and the grammar just under his coat, so that both could be retrieved in a hurry, raised himself and calmly awaited the Englishmen's movements.

But this was Antarctica and no call for hostilities! The chance meeting of two rival ships at the far end of the world appears, superficially at least, to have resolved into something of a mutual admiration society. Amundsen, who had been absent at Framheim, came skiing down the Barrier slope (no doubt his effortless progress well noted by the others) to extend greetings to *Terra Nova*'s company. Wilfred Bruce, Kathleen Scott's brother, was one of those invited aboard *Fram*.[10] He found the broad, stubby vessel that had carried Nansen on his Arctic voyage "very ugly outside" but within the diesel-powered ship was surprisingly neat, comfortable, and clean. Moreover, each of the men seemed to have his own cabin. Bruce, who exclaimed "curses all round" at the first sight of the Norwegian, now agreed that "all seemed charming men, even the perfidious Amundsen."

For his part, Amundsen found the Englishmen "extraordinarily pleasant" after lunching on *Terra Nova* with Nilsen and Lieutenant Kristian Prestud. Later in the day, *Terra Nova* resumed its westward journey with Bruce finally noting that the Norwegians "were very friendly, but didn't give away very much or get much." Privately, no one could ignore the fact that from what they had seen, Amundsen and his team were much more skilled in the arts of polar travel and much more prepared—especially in their complete reliance on dogs—than the force at Cape Evans. More, the Bay of Whales was just that much closer to the Pole. Scott would have to await his return to Cape Evans from the southern depot-laying party to learn of Amundsen's presence. Anticipating this moment, Bruce wrote that there could be "nothing so exciting as the *Fram*'s discovery, an eruption of Erebus would fall flat after that."

For Scott's expedition, a heartbreak moment of 17 January 1912 as they surveyed the tent and flag left at the South Pole by the Norwegians who had been there a month before them.

But *Terra Nova* was not the only surprise visitor to the Bay of Whales. Finding a group of Japanese occupying their lookout tent at the ice edge caused a certain degree of disquiet among the Norwegians. "We were a little surprised to see a vessel come in," wrote Nilsen on 16 January 1912. "Finally we saw the Japanese flag. I had no idea that the expedition was on the way again." Lieutenant Nobu Shirase led the first Imperial expedition on the 200-ton *Kainan Maru* ("opener of the south") reaching the Barrier after ice conditions forced a winter's frustrating delay in Sydney.[11] The two parties conversed, but with language difficulties their meeting was brief and the South Pole did not seem to be an issue. Nilsen estimated that about twenty-seven men were aboard *Kainan Maru*, including two Ainu tribesmen to drive the twenty-seven or so dogs. "It is a small, extremely dirty vessel . . .

The first Imperial Japanese Antarctic Expedition, led by Lt. Nobu Shirase, spent the 1911 winter in Sydney Harbor until Kainan Maru *could sail again for the Bay of Whales. On 28 January 1912, Shirase's "dash patrol" reached 80°S, where they gave three* Banzai! *for the Emperor.*

everything seems to be very disorganized . . . they are really quite wild."

With the aim of sledging "farthest south" into King Edward VII Land, Shirase's "dash patrol" on 28 January reached 80°5'S, a distance of 160 miles (260 km), where they flew the flag of the rising sun and gave three *Banzai!* for the Emperor. Among their modest achievements, the Japanese were the first to reach King Edward VII Land across the pack ice, where the *Discovery*, *Nimrod*, and *Terra Nova* voyages had failed in previous years; farther to the east of the Bay of Whales is found Kainan Bay, an inlet named in honour of their little ship. When the next expedition to the Bay of Whales returned fifteen years later, not a trace of Norwegian occupation would be found. Inevitably the sides of the Bay had calved and on an iceberg, Framheim had drifted out to sea.

Nor will any trace of *Kapitan Khlebnikov*'s visitation be found. A new regime of tidiness prevails in Antarctica, no longer a toleration of the mess that the old explorers spread around the huts at Capes Royds, Evans, or Adare. When we pause at Cape Washington for afternoon tea on the ice, every last cup and plate must be plucked from the snows, every napkin carefully dropped in the bin. At least we are at one with Amundsen; no one will find a trace of our coming.

Kapitan Khlebnikov is steadily pushing south . . . 76° . . . 77° . . . the Global Positioning indicator reads. How much will *Khlebnikov* find that the Bay of Whales has changed? The calving of mighty B15 may have totally redrawn the Barrier chart that previous explorers knew. Maybe strange ice territory awaits at the end of the world before we can conclude our bid for "farthest south."

Above: Kapitan Khlebnikov *docked against the*
ice. "At least we are at one with Amundsen . . .
one will find a trace of our comin

Right: Penguins, like these ever-present Adelie found on the Ross Sea coast, epitomize the fascination with Antarctica's wild life.

CHAPTER THREE

LITTLE AMERICA

Fortuitous it was for the Americans that Sir Douglas Mawson, regarded by them as "the greatest living Antarctic explorer," paid a business visit to New Zealand in November 1928. For Commander Byrd it presented a rare opportunity to meet with the Australian leader who had actually been to that remote part of Antarctica that was his very destination—the Bay of Whales.

As a young geologist from University of Sydney with Shackleton's expedition aboard *Nimrod* in 1908, Mawson had cruised the Great Ice Barrier and inspected the Bay at close quarters. A few years later, leading the

Australian Antarctic Expedition of 1911, he had suffered the loss of his two sledging companions in the depths of the Antarctic wilderness and barely survived himself. For Byrd, about to take his expedition south and intending to fly an aircraft to the Pole, the opportunity of a meeting with Mawson was not to be missed. A photograph taken at a Wellington office shows the two men together—Mawson in his business suit and Byrd in naval uniform, cradling Igloo, his pet fox terrier.

At 3:29 P.M. on Thursday, 28 November 1929, Commander Richard E. Byrd, USN, took off from his Little America base beside the Bay of Whales and headed towards the bottom of the world. Along the strip dug from the snow, forty men cheered as the expedition's Ford Tri-Motor laboriously climbed above the ice. Half of them must wonder if they would ever again see Byrd and his crew of three. The other half were sure they wouldn't. The first flight to the South Pole was a touch-and-go mission.

The mid-afternoon departure brought the sun abeam of the aircraft for both outward and return journeys, allowing sights for true heading to be taken through the cabin side windows. Byrd appointed

Right: *Commander Richard E. Byrd goes before the camera in a New York studio, wearing his full polar flying rig and holding the snowshoes he will wear in event of a forced landing.*
Below: City of New York, *one of the Byrd expedition's veteran ships, unloads its cargo at the Bay of Whales.*

himself flight commander and navigator, relying on a drift indicator and mariner's sextant with compensating bubble-level to locate the Pole. Begrudgingly he had chosen his chief pilot, the twenty-eight-year-old Bernt Balchen, to fly *Floyd Bennett*. The two did not much like one another but Byrd knew the stocky, powerful Norwegian, a naval aviator and internationally ranked boxer and skier, had the coolest and most experienced hand of any of his airmen—and coolness and experience were the qualities he needed that day.[1] Harold June, deputy to Balchen, was radio operator and general crew, a duty which included operating the Paramount newsreel camera. Ashley McKinley, ex-army photo-survey expert, managed the heavy Fairchild K5 mapping camera aimed through one of the celluloid windows of the stripped and packed passenger compartment.

They were 400 miles (640 km) across the Ross Ice Shelf when Balchen sighted Larry Gould's dog sledge team.[2] Gould, the expedition's second in command, at noon had radioed the clinching message: "Weather clear above the Queen Maud Mountains." They zoomed lower to drop chocolates, cigarettes, and photo charts to the waving men who had taken four weeks to travel the distance they had covered in four hours.

With the sun on their left shoulder, to the far right they began to see the peaks of the Trans-Antarctic Mountains glinting at them like burning glass. At another 100 miles (160 km) out, Balchen turned *Floyd Bennett* more sharply to the west. Two glaciers were then options in striving for the polar plateau, both mapped and named by Amundsen; one was the Axel Heiberg, honouring his backer; Liv was the other, for Fridtjof Nansen's daughter. The Heiberg crested with the plateau at around 10,500 feet (3,200 m), the Liv at about 1,000 feet (300 m) lower. "Take the Liv," shouted Byrd.

The big aircraft of the long blue fuselage and bright orange wing shrunk to a merest dot against the overpowering mountains that crowded

Above: Preparing Byrd's "flying washboard"—the big Ford Tri-Motor, Floyd Bennett*—for the South Pole flight from Little America in November 1929. Below: For Australian Post Office's Antarctic Territory stamps, artist Ray Honisett illustrated Admiral Byrd's moment of triumph on 29 November 1929—the Ford Tri-Motor in flight and, above the Pole, dropping an American flag attached to a stone from his friend's grave.*

the entrance to the Liv. They came in at 8,200 feet (2,600 m), Balchen urging the motors at full throttle. Ahead of them the glacier slanted mercilessly upward in a chaos of gigantic icefall and crevasse-scarred snowfield. Towards the southern end, where the heavily laden aircraft panted at a bare 80 mph (128 kph) their path became even steeper. For every length *Floyd Bennett* lifted its nose, a torrent of headwind from the plateau seemed to drop it again, almost as far. At least 1,500 feet (500 m) stood between them and the final crest; white-flanked mountains closed in, the glacier grew narrower; hardly space to turn, no alternative but to press on, in an aircraft about as responsive as a bucketful of lead.

From the cabin, looking forward through the cockpit window at a solid wall of ice Byrd described what followed:

> Balchen began to yell and gesticulate, and it was hard to catch . . . the words in the roar of the engines echoing from the cliffs on either side. But the meaning was manifest: "Overboard—overboard—200 pounds."
>
> Which would it be—gasoline or food?
>
> If gasoline, I thought, we might as well stop there and turn back. We could never get back to the base from the Pole. If food, the lives of all of us would be jeopardized in the event of a forced landing. Was that fair to McKinley, Balchen, and June? It really took only moment to reach the decision. The Pole, after all, was our objective. I knew the character of the three men. They were not so lightly to be turned aside. McKinley, in fact, had already hauled one of the food bags to the trapdoor.[3]

Byrd nodded and the precious provisions went splattering to the glacier floor. The straining machine lifted, but still not climbing fast enough against the relentless downdraft. They stared into another obstacle, the whitened hump of a mountain peak rising from the middle of the Liv; no way around it, only over the top.

"Drop more," the pilot shouted. "Another bag, unload!" McKinley grabbed a second sack and, with barely a glance at their leader, sent it plummeting; goodbye to enough rations for a month.

Like a man possessed, Balchen abruptly threw the bucking aircraft sideways, aiming them at the glacier wall, the right-hand wingtip close to scraping the unfriendly face of bare brown and blackened rock. No reply to their shouts of warning. His final gamble guessed at catching an eddy of the updraft that might lurk by the cliff on the very fringe of a ceaseless headwind. Suddenly a lightened *Floyd Bennett* roared skywards, rising on the current as if propelled by an unseen boot. To whoops of joy and relief from those in the cabin, they were over the crest with 450 feet (140 m) to spare. In that moment of terror, a cool head in the cockpit had saved them all.

The top of the Liv delivered them to the vast white desert of the polar plateau. The key peaks of the Queen Mauds fell away with Byrd's comment "the parade of mountains, the contrast of black and white, the troughs of the glaciers . . . something never to be forgotten." At about 1,000 feet (300 m) above the surface, the steady beat of the motors carried them through unbelievably low and oxygen-starved temperatures outside. Byrd had sometimes warned "any small failure could spell an end to it all, a flaw in a piece of metal, a bit of dirt in the fuel lines or in the carburetor jets."

Above the remotest place on the globe, Byrd's most dreaded moment struck; the aircraft began to vibrate with the spluttering and backfiring of the right-hand engine. June reached for the fuel dump lever, ready to drain the tanks to save them from a fireball if they were forced to ditch on the sastrugi ridges below. McKinley twisted the flow valve—this way, that way, trying to correct a possibly over-lean mixture before the stuttering propeller stopped altogether. Then another loud backfire and the erratic motor resumed its full-throated roar. All eased back in their seats, conscious of the slender margin between them and the cold hell below.

Another four hours of the eternal plateau: at 89°20'S, Balchen studied his slide rule and sent a scribbled note along the wire that connected to Byrd's position aft. "According to my reckoning we should be over the South Pole in fourteen minutes." "My reckoning agrees," Byrd replied, though Balchen privately wondered if their leader, with his inadequate sextant, really had any clue to where they were.

In the full daylight of 1:14 A.M. they made a broad circle above featureless ice. Byrd handed June his signal for transmission:

> Aboard Airplane *Floyd Bennett* in flight. 1:55 P.M. Greenwich time, Friday, November 29. My calculations indicate that we have reached the vicinity of the South Pole, flying high for a survey. The airplane is in good shape, crew all well. Will soon turn north. We can see an almost limitless polar plateau. Byrd.[4]

They dropped an American flag, weighed with a stone from the grave of the man who had been Byrd's pilot on the North Pole flight. Byrd's parting comment was probably one of the truest observations he ever made: "One gets there and that is all there is for the telling. It is the effort to get there that counts." They chose the Axel Heiberg for their return to the Ross Ice Shelf.[5] At the foot of the mountains awaited the cache of 1,500 liters (350 gallons) of gasoline which had been brought out aboard *Floyd Bennett* eleven days before. Balchen noted with some displeasure that while he and his two companions lugged the chill five-gallon drums to replenish near-empty tanks, Byrd, who had been swigging liberally from his cognac flask, chose to dance about yelling "We made it! We made it!"

After another four and a half hours they sighted the white arc of the Bay of Whales and then the landmark of Little America's three tall radio towers. At 10:08 A.M. they were climbing from the aircraft into the arms of a jubilant welcoming party. Hoisted on many shoulders, they were carried to the mess hall for a postponed Thanksgiving dinner. *Floyd*

Bennett had carried them 1,750 miles (2,800 km) in a flying time of 17 hours and 28 minutes, or a total absence of 18 hours and 41 minutes. During those hours, announced McKinley, he had exposed 1,600 frames on the K5 camera, covering an area of some150,000 square miles (388,000 square km) of Antarctica, much of it previously unseen. President Hoover greeted their feat as "proof that the spirit of great adventure still lives." Congress awarded the Medal of Honour and prepared legislation that would elevate Commander Richard E. Byrd to the rank of rear admiral (retired). In New York, a cheering crowd gathered in Times Square to watch the illuminated bulletin flash the news that for Thanksgiving Day, America had conquered the South Pole.

Admiral Byrd was the man who brought the mechanical age to Antarctica.[6] It was he who established the Little America legend on the Bay of Whales, where he sought to flex America's muscle in securing territorial rights to a major portion of the Antarctic continent. Just as Amundsen had followed the writings of men who went before him, Byrd absorbed what Amundsen had to say and made his decision in favour of settling at the Bay. The far eastern end of the Ross Ice Shelf had other advantages, too, being distant from the "British" sphere of occupation at McMurdo Sound and close to an unseen, unmapped region on which he had his sights set.

Here at the Bay he established his pioneering expeditions, Little America I of 1929, Little America II of 1934, and at his urging, Little America III of 1939. From Little America I he achieved the first flight to the South Pole and in the following expedition, taking wing to the east, saw and claimed for the United States the vast territory of Marie Byrd Land, which he had named after his wife. Camera-equipped aircraft and broadcast links with American radio stations would now be the norm. Byrd took 100 dogs on his first expedition but he also said "the sledge dog must give way to the aircraft, the old age is past."

Byrd's expedition with its eighty-three men, three aircraft, and a motor tractor reached the Bay of Whales in January 1929. After reconnoitering the ice edge for a suitable landing, he decided to place Little America some seven miles north of Framheim—such was the change in the Bay's shape since Amundsen's departure seventeen years before. His ships were *City of New York,* a square-rigger built in 1883, and *Eleanor Bolling*, a battered former wartime mine sweeper of 800 tons.

Unloading beside the Barrier became a tricky business. In one risky operation, the overhanging cliff abruptly dropped an avalanche of ice on the deck of a near-capsizing *Eleanor Bolling;* when one sailor fell from the ice, Byrd himself was the first into the water to rescue the man. Frequently parts of the Barrier collapsed around them. Byrd reported "constantly hearing the reverberating, long-sustained echoes of the Barrier crumpling to the west and north, and the sharper, more piercing reports of splitting bay ice." Men's lives were constantly in danger as the world appeared to disintegrate around them—an expeditionary named Harrison hung by a slender rope as a cliff edge broke away, another named Roth was washed into the bay, clutching frantically at a fragment of a floe; yet despite the litany of near disasters, not a single man was lost nor, it seems, even seriously hurt.

On another occasion danger came from a different quarter in the Bay when Byrd's motor boat was pursued by a pod of about ten killer whales. "I do not consider myself a particularly imaginative man," wrote Byrd, "yet I confess candidly that the sudden appearance of these ill-reputed creatures had a disturbing effect: we became most sensitively aware of the flimsy character of our boat." Running at full speed, they reached the protection of an ice edge where, Byrd continued, "As we faced around the killers came up not more than fifteen feet from where we stood. Another dive would have brought them up with us. We had drawn our revolvers—a foolish gesture, I concluded later, for a battery of 75's would not have stopped them."

The $750,000 he had raised from some of America's biggest corporations and wealthiest industrialists had enabled Byrd to purchase three aircraft. All high-wing monoplanes, they were the Ford Tri-Motor and a single-engine Fokker and Fairchild; all were radio equipped. The big Ford, with a fuselage made of duraluminium (it earned the nickname of "the flying washboard") and with a total 975 horsepower, was the aircraft chosen for the Pole flight when the following summer arrived.

The weight of the investment in him by the Fords, Rockefellers, and Guggenheims caused Byrd months of worrying that the Australian aviator Sir Hubert Wilkins, flying from a base on the Antarctic Peninsula, would beat him to the Pole. "He wants to lick us," went the cable from Byrd to his agent in New York. "Don't forget Hearst, as rival newspaper magnate, offers him $50,000."

"The loneliest city in the world," as Byrd termed Little America I, with its scores of dogs and forty-three wintering-over expeditionaries buttoned up against the blizzards, listening to the bark of the departing Weddell seals and quietly praying not to awaken one morning and find themselves upon an iceberg adrift.

Men of many qualities, personalities, and outlooks inevitably comprised the Byrd expedition. But perhaps no single inhabitant of Little America was more complex and unpredictable than the leader himself. Monuments to Admiral Byrd stand in Washington, D.C.; at McMurdo Sound, the present-day American base; and in Wellington, Dunedin, and Christchurch, his New Zealand points of departure on the great southern adventure. Before the space age, he led the United States in the race to uncover the newest of worlds and unravel its ice-shrouded mysteries. To look at the map of Western Antarctica is to witness the extent of his exploits in the host of names that he left behind, all of them wholly American—and most of them names of family or businessmen closely associated with himself. As far as Americans were concerned, he dominated Antarctic exploration; over three decades, he mounted or led five expeditions and placed on the map more than two million square miles of the great white southland, a record on this globe difficult to duplicate. To his countrymen, he was indeed "the Lord of the Ice . . . Admiral of the Ends of The Earth."

New York showers a ticker tape welcome on Richard E. Byrd, America's explorer-hero of the 1920s. Lieutenant Byrd returned from his 1926 North Pole flight to be promoted to the rank of Commander; after the South Pole flight of 1929 he was made a Rear Admiral. But not all serving Navy officers were impressed at his rapid rise through the ranks.

It was his aim to establish United States sovereignty over his discovery of Marie Byrd Land, which stretched between the British and New Zealand sectors from longitudes 80°W to 150°W. At the limit of his sledging journey, Byrd's deputy, Dr. Larry Gould, deposited a paper which said "We . . . claim this land as a part of Marie Byrd Land, a possession of the United States of America." During Byrd's lifetime, the political climate in Washington changed and no official action has been taken to place the vast sector of ice territory under the United States flag.[7]

Indeed, the United States seemed to let no claim to sovereignty stand in its way upon a decision taken in far-off Washington to establish a major polar base for the International Geophysical Year. An extract from a story I wrote in March 1958 for the Australian and overseas press depicts an episode in the life of another "little America" at 77 degrees South.

> McMurdo Sound (Antarctica)—In sight of a hut that Captain Scott used on his last tragic dash to the Pole we are marooned in Antarctica waiting for a ship to take us out. No planes can get in and none can get out. Not 200 yards from the hut where I am writing this story the sea is coming back into McMurdo Sound. The great ice airfield, which only a few days ago held Globemaster transport planes, is breaking up and drifting away.

OFFICIAL CAR
1

Eighteen men living in blizzardly isolation of the South Pole, 800 miles from here, are now cut off for the winter. "Button up and do the best you can with what you have," Admiral George Dufek, leader of the United States Navy's "Operation Deep Freeze," has ordered them.

All day long a Navy helicopter has been buzzing overhead, lifting the last airfield equipment off the disintegrating ice runway as Lieut Commander Edward Ludeman and his construction chief, veteran Commander Roger Witherell, of the U.S. Seabees, act to beat the emergency. Two miles from here across a windswept pass the Americans are building their small alternative landing field as part of the Commonwealth Trans-Antarctic Expedition strip at New Zealand's Scott Base.

Already they have two silver Dakota aircraft and one small Otter cabin monoplane parked alongside the planes of the Expedition. From here the Navy's jet-assisted Neptune bombers, now being held at Christchurch, New Zealand, would take off for the Pole if a matter of life and death arose—but only when Admiral Dufek has given the order.

Not so far in the experience of Operation Deep Freeze has the ice in McMurdo Sound been known to break up so early, so suddenly and so treacherously. Half an hour before the last parachute supply drop of the season was due to leave for the Pole the alarm was given that the ice was going out. By loudspeakers all seventy outgoing members of the first summer party and International Geophysical Year scientists were ordered to board the waiting Globemaster. The scenes that followed were reminiscent of the best traditions of wartime movies as men streamed out of huts hauling baggage, zipping up clothes, pulling on hats, throwing gear aboard revving trucks and hurrying to the helicopter take-off spot. For two hours two Navy "choppers" loaded with men and freight made non-stop shuttle trips to fill the belly of the Globemaster standing out on the ice.

Finally the huge aircraft lumbered off, made a deafening run over the now silent base and passed out north towards New Zealand and civilization. The next day the Navy's largest icebreaker, *USS Glacier,* loaded another eighty men of a second priority outgoing summer party from the end of the shattered ice roadway that once linked with the airfield and churned northwards towards the relief of other bases at Cape Hallett and Little America on the Ross Sea, then back to New Zealand.

It is growing lonely in McMurdo now—the place they call the Crossroads of Antarctica—biggest and one of the most southerly bases in all the polar continent. Last week the Antarctic sun set for the first time this season and snow flakes drifted on the base from the slopes of 12,000-foot Mount Erebus, the only active volcano in Antarctica, which is but thirty miles from here. A husky is howling outside the door tonight; the barking of seals comes faintly on the wind, and I hear the sound of crunch and crumble as hundreds of tons of ice fracture and float away on the Ross Sea.

Life at McMurdo, three thousand miles from Sydney, has compensations—a town of odd habits and strange

The cook at Scott Base contemplates a "deep frozen" dinner. U.S. Navy aircraft stand on the New Zealand strip, evacuated in the 1958 ice break-out.

contrasts. It is a place that seems surrounded by all the snow in creation yet where they have water rationing in summer and showers are restricted to once a fortnight.

It is a place where you can eat four meals a day—the last at midnight—and where they have to close the shutters before beginning the second session of the movies at 1:30 in the morning so that the glare of the midnight sun will not spoil the screen. It is a town where the clink of money is unknown, for everything is on credit—where you can only obtain beer by the caseful (American beer in cans), but where hard liquor is forbidden.

It is a place where refrigeration is needed to keep food hot and rain is unknown and where the air is so dry that you can still sample biscuits Scott left in an open box outside his depot at Hut Point, half-a-mile from here, forty-six years ago.

I was the last correspondent to get into Antarctica before this winter and I will be in the last party to get out—I hope.

Richard Evelyn Byrd, who came from an old Southern family, was born at Winchester, Virginia, on 25 October 1888. At the age of twelve he was sufficiently mature to have his parents send him alone by ship to visit friends in the Philippines. At twenty-four, he joined the United States Navy, only to be invalided out four years later because of a recurrent leg injury. World War I brought his recall to the mast, except for him it was to be a desk in Washington until he practically browbeat his superiors into letting him train for a combat role where not so much store was put on strong ankles. He entered the Pensacola flying school in Florida and, as hostilities ended, obtained his wings.

Byrd emerged from his wartime training with exploration as his goal; to make a name for himself as an aviator who rolled back the curtain of mystery above far-away places. His early ventures included flying the Atlantic, though mishaps robbed his chance of beating Charles Lindbergh's history-making crossing; out of fuel and lost in darkness, his plane crashed-landed on the French coast.

Backed with a $100,000 fund provided by his friend, Edsel Ford, he set out to be the first man to fly an airplane to the top of the world. With Floyd Bennett, an experienced Arctic aviator, as his pilot and Byrd himself filling the role of navigator, he claimed the North Pole on 9 May 1926. An instant American hero, he was promoted by order of Congress from lieutenant to commander (he is said to have refused rear admiral as too much of a leap). New York welcomed him home with a ticker-tape parade.

A meeting with Roald Amundsen at Spitzbergen airstrip on return from the North Pole flight may have determined his next path to glory. "Where to now, Byrd?" asked the Norwegian explorer. Almost involuntarily, Byrd replied "the South Pole, of course." Richard Evelyn Byrd's place is secure as one of the foremost explorer-leaders in the twentieth century's new age of technology. He was a gifted organizer, had access to the corridors of power, be it politics or commerce (his brother was Senator Harry Flood Byrd), and for him, thinking big was as natural and American as apple pie. His organization brought an enormous advance to polar discovery in the use of aircraft, radio, and tracked vehicles. His other attribute was a superb sense of showmanship. He knew how to employ public relations to make the American populace sit up and take notice; he signed contracts for books, magazine articles, newsreels, radio, and personal appearances.

The same R. E. Byrd was indeed a complex individual; some of his critics would condemn him as a publicity-grasping figurehead, propped up by influence and the labours of other men. Among his expedition members were those who would refer to him as jealous, divisive, suspicious, ill-tempered, and where flying and navigation were concerned, of questionable ability. In his own service, because of his rapid and "political" rise to two-star rank, he was never highly popular. Russell Owen, the correspondent for the *New York Times*—which was paying $60,000 for an exclusive service on the 1929 expedition—is reputed to have lapsed into a deep depression and retreated to his bunk for several days when Byrd demanded the right to censor and change his newspaper copy. Shortly after the first expedition reached Little America, Byrd aroused the hostility of his dog team leader, whose chain of authority he arbitrarily altered.

Stories such as these, particularly centred on the first Little America expedition, were recounted by Norwegian-born Bernt Balchen, whose relationship with his leader appeared to have been far less than easy. Though Balchen was chief pilot, other airmen were privately told by Byrd that they would be his preference for the prized South Pole flight. Balchen complained of Byrd's utter aloofness towards him, yet he was the one on whom Byrd ultimately depended to get *Floyd Bennett* to the Pole. On that epic flight, said Balchen, it was he who had to tell the leader when they were nearing the Pole, for Byrd himself had lost his sense of direction. For his part Byrd once admitted, almost in annoyance, that Balchen "can do more things well than any man I ever knew."

Balchen also accused Byrd of having wilfully sewn the seeds of jealousy among the mechanics who were entrusted with the vital task of aircraft maintenance. Each one of them was apparently taken aside by the leader and told that he alone had Byrd's confidence and that the others should be severely monitored. The incident led to bad blood in the workshop until detected by Balchen, who had no wish to fly across Antarctica in an airplane fixed by disaffected men.

Strangest of all was Byrd's decision at Little America II to subject himself to nineteen weeks of isolation in what was called the Bolling Advance Weather Station, located 125 miles (200 km) south of the Bay of Whales. Byrd proposed to transmit daily weather reports from a tiny hut, buried in the ice. His lone vigil began on 24 March 1934, but within a few months the stumbling and partly incoherent radio signals made it obvious that something was seriously amiss. An oil stove with a faulty flue and a petrol generator leaking fumes were, in fact, poisoning Byrd, who could not turn off either when the outside temperature hovered around –50°C (–69°F).

On 11 August, after several futile attempts, a three-man party set out in a Citroen tracked vehicle to make a hazardous drive through the darkness. At the advance base they found their leader emaciated, debilitated, and close to death from carbon monoxide fumes. They remained with Byrd for another two months before he was well enough to return to Little America, where he underwent a further spell of recovery. Strange behavior indeed.

Why Byrd would want to place himself in such utter isolation from the men he was supposed to lead remains hard to fathom. Byrd detailed the almost fatal experience in his book *Alone*, which became a bestseller. Recounting that he was prompted to maroon himself because of an "interest in the experience for its own sake" seemed to be another way of saying that by some strange quirk in his nature, he was forever trying to prove himself; that he was perhaps instinctively hostile to some of his rivals or subordinates for fear that their achievements would overshadow him in the hall of fame. One notes, for instance, that Byrd's writing never gave credit to Hubert Wilkins for the first flight in Antarctica. Indeed the Australian's exploits are barely mentioned. Though the late 1920s were prohibition time in America, Byrd ensured that the expedition stores contained an amount of liquor for medicinal reasons. "Booze" became a serious problem among some of the men at Little America, not helped by the leader's own drinking habits. On the Pole flight Balchen recounts that Byrd helped himself liberally to a flask of cognac he usually carried because of a supposed heart ailment. On the occasion of a seven-hour flight to the Queen Maud Range on 5 December, Balchen described the unseemly posture of the commander as the aircraft doors were opened at the return from the long journey. Assembled base personnel beheld their leader pinned to the floor by June and

McKinley, struggling and cursing loudly; in short, he seemed to be "raving mad."

Balchen himself could be temperamental and to some degree like natures must collide. Yet the reason for Byrd and Balchen's falling-out may have come from the darker side of exploration and its practitioners. According to Balchen, his friend Floyd Bennett—Byrd's pilot on the first epic flight that gained him national fame—confessed to him before his death from pneumonia in 1928 that, in truth, they had never reached the North Pole. Byrd's resentment may have been directed at the man who knew his secret—if indeed such a secret did exist.[8] Again, that other American polar leader, Finn Ronne, told much the same story which disputed the admiral's moment of glory. In more recent years the success of that flight has been seriously questioned.

Byrd may have been a strange character, audacious, theatrical, and ambitious. Yet he was generally careful for the welfare of his men, to which the record of few accidents and major injuries on his expeditions is testimony. He also showed himself to be divisive in his dealings with those he placed in authority and unpredictably moody. Perhaps his erratic nature embodied those characteristics that are sometimes inseparable from an indefinable charisma, the strange stuff of leadership. Certain of Byrd's biographers have suggested that the admiral, despite his achievements, was a "white knuckle aviator" and nervous of flying. But aviation is inseparable from travel, exploration, and tourism in modern Antarctica. In the almost seventy-five years since that remarkable Australian adventurer Sir Hubert Wilkins pioneered Antarctic aviation, the aircraft has accomplished what dog sledge or tracked vehicle could never remotely attempt.

Approaching the Bay of Whales reminds me that I, like Admiral Byrd, have made a first flight to the South Pole, though my claim perforce must be much more low key. Back in 1964 I was one of the two Australians privileged to be invited by the United States Navy to participate in the first (and so far only) flight from Australia to the Pole. Our ski-equipped C-130 Hercules took off from Melbourne and was aloft for 15½ hours and covered 4,420 miles (7,110 km).

McMurdo, enveloped in a blizzard, was unavailable and our plane was forced to divert to Byrd Station where, with the cabin depressurized, nose landing gear inoperable, and out of fuel, we made our touchdown under crash landing conditions. At Byrd Station, located at 80° S, and an elevation of some 5,500 feet on the Rockefeller Plateau of Marie Byrd Land, we lived in tunnels beneath the ice until a relief aircraft arrived. For me it was the second time at Byrd, the first being in the early years of Deep Freeze when we stayed (as I reported to the newspapers) in the original base that was fast sinking under a monstrous burden of blizzard-blown snow; above our heads the raw ice ceiling sagged around a bunch of oil drum supports that looked like so many squashed top hats. Not a good recipe for sound slumber.[9]

> The visit to "new" Byrd Station began with a descent through the snow in a tunnel reminiscent of illustrations of the entrance to Dante's Inferno that scared me as a child. But no inferno awaited below—instead a wide and silent floodlit roadway enclosed by smooth walls and roof of blue-tinged ice. Three observation towers protruding through the ice cap, a web of aerials and an exhaust vent, plus a 12,000-foot runway bulldozed across the plateau were the only clues to our existence below.
>
> To build a new Byrd Station the Navy brought in the Seabees—sixty-five men of the Construction Battalion who had worked previously on Greenland's Camp Century. With them came two powerful Swiss-made ploughs, known as Peter Snow Millers, which, as I watched, sliced through the ice at an amazing rate (dare I say "like cheese"), gradually making a new series of trenches for a base extension.

From a distance the Snow Millers looked like erupting geysers as they spouted a plume of finely shredded snow and ice high into the air. The trenches, eventually reaching thirty feet deep, are covered with steel arch beams on which the milled snow is resettled into a hard concrete-like surface. Across them the Antarctic blizzards can harmlessly swirl and scream.

Byrd is exclusively given to scientific observation. In these calm, protected quarters men stay for a year at a time to study meteorology, geomagnetism, glaciology, aurora, gravity, radio noise (the eerie "whistler effect"), the ionosphere and air glow. Like a normal township, Byrd is arranged around a central "main street" linked to seven smaller "side streets" and numerous interconnecting passageways. One tunnel houses the comfortable prefabricated huts containing the dining room, kitchen, and recreation facilities; another has the administration and communications center; a third has the medical quarters and various laboratories; yet another has the powerhouse, garage and workshop—and so it goes on. And every single section, down to the last bolt and screw, came in by airdrop or ski-equipped Hercules and over-ice tractor train.

The temperature in the tunnels remains at a steady few degrees below zero while within the insulated huts it could be shirt sleeves on a mild summer's day. Strain gauges attached to the tunnel walls detect changes in the ever-restless ice cap; they are a timely reminder of the gigantic forces surrounding us that might one day put continued occupation of the Station at risk.

Meanwhile, as the polar sky grows black and blizzardly, the troglodytes of Byrd can "shut the trapdoor" above their heads and get on with studying the brilliance of the aurora and listening to the weird mutterings of the heavens, undisturbed by the storms that rage beyond the entrance tunnel. But it's vital to obey the rules; a visiting Japanese scientist who ventured outside in darkness was never seen again.

Two *Mi* helicopters from Vladivostok Air are carried on the rear deck of *Kapitan Khlebnikov.* Their mission is to fly ahead of us to locate soft ice or open-water leads and to provide the link between ship and shore, as when we flew into the Dry Valley or visited Cape Evans and McMurdo base.

Many icebergs are visible from *Khlebnikov*'s bridge. The helicopters are warming up. Admiral Byrd would like this.

Southern Cross

Almost sixty years after Captain James Clark Ross noted a break in the Great Ice Barrier, the *Southern Cross* brought the Norwegian-born Carstens Borchgrevink and his Australian expedition to the same elusive location.[2] On 16 February 1900, Borchgrevink recorded his impressions of finding a "large oval basin" which he decided to risk investigating:

Southern Cross anchored at Cape Adare.

"On the 16th we were still proceeding southwards with plenty of "pancake" ice around us. On that date I discovered a break in the barrier, with low ice towards the east. The break appeared in two conspicuous heads. They were about one mile apart, opening up into a large oval basin, some four or five miles in diameter. Towards the west the barrier was about the same height as we had found it all the way from Mount Terror eastwards, but towards the south it started to fall, and round to the east it was quite low, only some two or three feet above the sea-level, rising gently towards south-east until it gained the normal height of the ice-sheet in the vicinity—some seventy feet. We entered, and reached lat. 78°34'S.

"Captain Jensen and I had inspected the appearance of this natural harbour thoroughly from the crow's nest before we decided to enter, fully recognizing the possible risk of being shut in in case the heads, through changes in the ice sheet, might close together. During the time we spent in this interesting harbour constant watch was kept in regard to movements and sounds in the ice. At a low place we moored the *Southern Cross* to the ice sheets by ropes and an ice-anchor. Here I effected a landing with sledges, dogs, instruments, and provisions, and while I left the sledge in charge of Captain Jensen with the rest of the Expedition, I myself, accompanied by Lieut. Colbeck and Savio, proceeded southwards, reaching 78°50', the farthest south ever reached by man."

Discovery

Inflating Eva at Balloon Bight in February 1902.

My flight of thirty-seven years ago would have taken us over the Bay of Whales at the position of "Balloon Bight," where Robert Falcon Scott became Antarctica's first "aeronaut" in *Eva,* a gas-filled Army balloon. Scott does not appear to have much enjoyed his brief though historic excursion into the polar sky. Having moored *Discovery* at the inlet which he named, instructions were given to Mr. Skelton, the ship's engineer, to commence inflating *Eva* from nineteen cylinders the Army had provided, each holding 500 cubic feet of gas.[3] On 4 February 1902, Scott recorded his impression as the first man to gaze down upon the face of Antarctica:

"The honour of being the first aeronaut to make an ascent in the Antarctic Regions, perhaps somewhat selfishly, I chose for myself, and I may further confess that in so doing I was contem-plating the first ascent I had made in any region and as I swayed about in what appeared a very inadequate basket and gazed down on the rapidly diminishing figures below I felt some doubt as to whether I had been wise in my choice.

"Meanwhile the balloon continued to rise as the wire rope attached to it was eased, until at a height of about 500 feet it was brought to rest by the weight of the rope; I heard the word "sand" borne up from below and remembered the bags at my feet; the correct way to obtain greater buoyancy would have been gradually to empty these over the side of the car, but with thoughtless inexperience I seized them wholesale and flung them out, with the result that the *Eva* shot up suddenly, and as the rope tightened commenced to oscillate in a manner that was not at all pleasing. Then, as the rope was slackened I again ascended, but, alas! Only to be again checked by the weight of rope at something under 800 feet. When I again descended to the plain, Shackleton took my place, armed with a camera. I had hoped that in the afternoon other officers and men would have been able to ascend, and especially our engineer, Mr. Skelton, and those of his department who had so successfully inflated the balloon, but the wind was gradually increasing, and our captive began to sway about and tug so persistently at its moorings that it became necessary to deflate it."

Fearing enough risks had been taken, Scott terminated the flying program and *Eva* packed away for keeps. One cannot but wonder how different the annals of Antarctic exploration might be, had the ropes parted and Scott shot heavenwards, never to be seen again?

Nimrod

Scott, it may be recalled, had extracted an undertaking from Shackleton that he would not settle on McMurdo Sound, where the Royal Navy captain insisted he had prior rights. On his first voyage, Shackleton contracted scurvy while on a sledging mission with Scott and had been sent home when *Discovery* sailed to avoid the winter ice. Clearly there was a reflection that he was "not up to it," for Scott tended to regard illness as a sign of weakness. Courage and determination brought

Shackleton back to Antarctica with his own vessel *Nimrod* and his remarkable British Expedition of 1907–1909—which explains why early in 1908 he was cruising the Barrier, searching for a landing place to honour a very questionable "gentleman's agreement." Beside him were the young Australian geologist Douglas Mawson and Professor Edgeworth David when he gave the order to turn the ship into a half-obscured harbour in the Barrier front—on which he would also bestow a name.[4]

"About midnight we suddenly came to the end of a very high portion of the Barrier, and found as we followed round that we were entering a wide shallow bay. This must have been the inlet where Borchgrevink landed in 1900, but it had greatly changed since that time. He describes the bay as being a fairly narrow inlet. On our way east in the *Discovery* in 1902 we passed an inlet somewhat similar, but we did not see the western end as it was obscured by fog at the time. There seemed to be no doubt that the Barrier had broken away at the entrance of this bay or inlet, and so had made it much wider and less deep than it was in previous years.

"The heavy ice and bergs lying to the northward of us were setting down into the bay, and I saw that, if we were not to be beset, it would be necessary to get away at once. All round us were numbers of great whales showing their dorsal fins as they occasionally sounded, and on the edge of the bay-ice Emperor penguins stood lazily observing us. We named this place the Bay of Whales, for it was a veritable playground for those monsters.

"We coasted eastward along the wall of ice, always on the look-out for the inlet. The lashings had been taken off the motor-car, and the tackle rigged to hoist it out directly we got alongside the ice-foot, to which the *Discovery* had been moored; for in Barrier Inlet we proposed to place our winter quarters. However, the best-laid schemes often prove impracticable in polar exploration, and within a few hours our plan was found impossible of fulfillment. The Inlet had disappeared, owing to miles of the Barrier having calved away, leaving a long wide bay joining up with Borchgrevink's Inlet, and the whole was now merged into what we had called the Bay of Whales."

If Antarctica gained a new place name, in the next breath one might say, Shackleton also lost a base. A landing place to

Shackelton's Nimrod pauses briefly beside the Great Ice Barrier.

establish the winter quarters from where he would march to the South Pole was wholly impossible. Instead of the convenient ice edge, which he recalled from 1902, a rampart of menacing cliffs where he dare not settle overhung the Bay. In his notes of 24 January 1908, Shackleton continued:

"This was a great disappointment to us, but we were thankful that the Barrier had broken away before we had made our camp on it. It was bad enough to try and make for a port that had been wiped off the face of the earth, when all the intending inhabitants were safe on board the ship, but it would have been infinitely worse if we had landed there and that when the ship returned she should find the place gone. The thought of what might have been made me decide then and there that, under no circumstances, would I winter on the Barrier, and that wherever we did land we would secure a solid rock foundation for our home."

Ironically, four years later it was left to Roald Amundsen to observe "if Shackleton had stayed on the Bay of Whales, we might not be the first people to have reached the South Pole."

Drifting outriders of the Antarctic continent, icebergs of many shapes, sizes, and electric colors await the southern voyager.

CHAPTER FOUR

Beaten to the Bay

On 19 January 1936, the Australian explorer Sir Hubert Wilkins sailed into the Bay of Whales in search of a missing millionaire. Wilkins, after Sir Douglas Mawson regarded as Australia's foremost polar explorer, was strictly following instructions. Lincoln Ellsworth, his American employer, had left Dundee Island almost two months before to make the first aerial crossing of the Antarctic continent.

In *Polar Star*, a Northrop low-wing monoplane, and with Englishman Herbert Hollick-Kenyon as his pilot, Ellsworth's flight plan

Left: Sir Hubert Wilkins' own polar exploits made him a considerable figure of world renown. In New York, Wilkins and his vivacious wife Suzanne unveil an ice sculpture at a Ford Motor Company dinner. Lady Wilkins was an actress, the former Suzanne Bennett, born at Walhalla in Victoria. She became a recognized portrait painter in New York.

led from the Antarctic Peninsula to Byrd's now abandoned Little America II camp at the Bay of Whales, a distance of some 2500 miles (4000 km). At the Bay, the pair intended to await Wilkins' arrival in the expedition vessel *Wyatt Earp* to pick them up and take them home to New York. Simple, eh?

But eight hours into their journey, *Polar Star*'s radio suddenly ceased transmitting and despite much monitoring of the airwaves, the radio remained silent. Newspapers soon headlined Ellsworth's disappearance. Mary Louise Ellsworth summoned her husband's wealthy New York friends to see what could be done. Gloomy press stories began to suggest that the millionaire adventurer and his pilot were surely dead. But Wilkins had supreme faith in their flight plan and its various emergency alternatives. Through a month's silence he calmly waited at the Peninsula for a message, until the moment came when he knew it was time to sail across stormy seas for a rendezvous beside the Great Ice Barrier.

Wilkins was no stranger to the Bay of Whales. Two years before, he and Ellsworth had attempted a "west-east" transcontinental flight from the Bay across to the Weddell Sea.[1] Sailing the 400-ton *Wyatt Earp* from Dunedin, New Zealand, on 10 December 1933, their original plan seemed smooth enough. Bernt Balchen, the now distinguished Norwegian airman, was their pilot. *Polar Star,* the latest all-metal monoplane from Jack Northrop's works in California, was the aircraft.

Punching for 450 miles (720 km) through the Ross Sea pack, they reached open water early in January 1934 and safely entered the Bay of Whales. Preparations were begun for Balchen's first flight on skis from the surrounding ice. Perhaps the entry in second mate Olsen's diary should have warned them that all was not well:

> Suddenly, from within the deep caverns far below the surface of the ice barrier, came ominous sounds like the tuning up of a mighty orchestra. It was as if the whole universe itself had begun to vibrate. We all stood rooted to the spot, so great was the terror which engulfed us. As the ominous sounds continued, it was as if a mammoth organ began to play accompanied by gigantic "cymbal" crashes . . .

Balchen and Ellsworth took *Polar Star* on a trial flight lasting thirty minutes on 12 January. They parked the machine overnight beside the ship on a gently rocking floe and all retired to the sound of soft music from Wilkins' gramophone. In the early hours of the morning the watch shouted for them to come on deck. In the grey light they beheld the floe split apart and their precious aircraft suspended above the sea by only its wing tips. Six hours of all hands labouring with ropes and winches brought *Polar Star* to safety. The Bay of Whales had sounded its warning note, and they had not listened. Before it really started, the 1934 Antarctic summer flying program was ended.[2]

The following year, with Bernt Balchen once again engaged as pilot, Ellsworth renewed his plan to make the pioneering long distance Antarctic flight. However upon Wilkins' advice—the Australian was still Ellsworth's faithful manager—the direction of travel went into reverse, starting from the Peninsula and heading west.[3] After several disappointments, a good flying surface was found at Snow Hill Island and on 3 January 1935, Ellsworth turned to Balchen and, as if almost on impulse, exclaimed—"Let's make a try! What do you say?" Before the take-off, an excited Ellsworth dictated a message to the North American press: "Flash—Balchen and I took off at seven this evening, heading for the unknown. The great adventure so long awaited is at hand."

A sponsor's product is well in evidence as Sir Hubert Wilkins (minus beard) and his aviation crew prepare to sail for the Antarctic Peninsula from Montevideo in 1928. From left: Joe Crosson (pilot), Wilkins (leader), Carl Ben Eilson (chief pilot), Orval Porter (mechanic).

Unfortunately, after only a couple of hours the "great adventure" came to a sudden halt when Balchen turned back because of looming bad weather. Ellsworth was furious. Balchen's parting remark was addressed to Wilkins—"Ellsworth can commit suicide if he likes, but he won't take me with him." On one of these occasions of failure and setback, the normally serene Wilkins, son of a South Australian sheep farmer, was prompted to exclaim—"If I'd have known this was going to be a one-day flight, I would never have joined the expedition!"

Wilkins himself was a polar explorer who had no need to apologize. Of him it was once said "he feeds on adventure as a child does on mother's milk." Extraordinary deeds of gallantry earned Wilkins the Military Cross and Bar during the War. General Monash, a famous Australian commander of World War I, called him "the bravest man I've ever met!"

Qualified as a pilot and skilled in navigation and photography, he made his mark as an explorer in attempting the first North Pole flight, only to be beaten by Byrd. In 1931 he tried but failed in a bid to reach the Pole in a pensioned-off wartime submarine. His knighthood came from the major Arctic flight between Point Barrow in Alaska and Spitzgen in Norway, when he and pilot Carl Ben Eilson were aloft for twenty hours and twenty minutes.

Like other North Polar men, Wilkins' adventurous spirit next heard the call of Antarctica. Raising a small expedition, and with financial support from the Californian press magnate William Randolph Hearst, he purchased two Vega high-wing Lockheed aircraft and invited Ben Eilson to be his pilot. From the whaling station at Deception Island on 16 November 1928 they made the first airplane flight above Antarctica. Five days before Christmas, they completed a ten-hour flight across the Peninsula's unknown mountains, during which an exuberant Wilkins wrote, "[F]or the first time in history, new land was being discovered from the air . . . I felt liberated . . . I had a tremendous sensation of power and freedom." A second expedition returned to Deception Island in summer of the following year, completing more aerial exploration of the Peninsula between December and the February of 1930.[4]

If Wilkins' exploits had carried Antarctica into the air age, his very presence on the far side of the continent late in 1929 caused much anguish at Little America, where a nervous Commander Byrd lay awake wondering whether "the Australian" would be first to claim South Pole honours. Byrd's paranoid fear of being beaten is documented in the cable messages sent to Hilton Railey, his agent and fund raiser in New York:

> You must not forget that Wilkins is out to lick us. I wish to impress upon you that the flight he proposes is even more important than a flight to the south pole. He is flying over the area that we are most anxious to explore and which is most important to science. As he starts much farther north than we do, he can start early and he is going to make every effort to beat us to it. Don't forget that he was offered $50,000 by Hearst to beat us to the south pole and that he will now possibly fly here by way of the south pole. In spite of this we have got to be sports and have got to be square with him, but do not give any information as to when we start flying. If he thinks we are going to start early he will naturally hurry the more . . . We must be very careful not to do anything that is lacking in sportsmanship.[5]

But Balchen's skill and *Floyd Bennett*'s endurance brought Byrd victory at the Pole, and in *Little America*, his book of the expedition, Wilkins' name is mentioned only once.

Wilkins was aged forty-five and Ellsworth was fifty-two when they met in a New York club in the summer of 1932 to discuss an Antarctic adventure. As a professional explorer, Wilkins (for those who could afford his advice) was a highly valuable and rare commodity, while Ellsworth (who could certainly afford him) belonged to that other exalted class for whom money is no object.[6] In Ellsworth's case it came from a family coal-mining fortune. No armchair adventurer, Ellsworth, as a graduate engineer, had worked on railroad construction in Canada and probed the jungles of South America; some might regard him as slightly eccentric, the particular hero of his life being Wyatt Earp, the gun-fighting lawman of the Wild West. He wore Wyatt Earp's wedding ring, hung Earp's gun belt in his cabin and his very ship he had renamed in the sheriff's honour.

"I would not be defeated," Ellsworth once declared when all the odds seemed against him and friends suggested a rocking chair. "I will cross Antarctica by air." For him the moment of opportunity at last appeared to arise on that 22 November in 1935 when a heavily laden *Polar Star* took off from the snows of Dundee Island and turned south. They were equipped to fly across thousands of miles of mountains and plateau, landing when necessary to rest or dodge storms, until reaching Marie Byrd Land, where, by swinging north, the flight would terminate beside the Bay of Whales; in an empty but well provisioned Little America II they would rest until Wilkins arrived to take them home. It was a plan that seemed very logical and simple if one discounted the fact that this was a journey across a wild, hostile, crevasse-ridden, blizzard-bound continent larger than the United States—and in all the interior of Antarctica, Ellsworth and Hollick-Kenyon were probably the only two men abroad. And now it had swallowed them.

After a month of waiting at Deception Island and receiving no signal from any of the optional landing places, Wilkins knew he must put the next part of the plan into action. On December 22, and with a relief Gamma 2 aircraft loaded in Chile, the anchor came up and *Wyatt Earp*, a former Norwegian herring trawler, began the long journey towards the Ross Sea. Ellsworth would surely be waiting for him—such was Wilkins' confidence in his employer's determination and his faith in the quiet Englishman who was now their pilot.

But the scene presented to him on rounding a headland of the Great Ice Barrier was like a ghostly replay of the Scott expedition's meeting with Amundsen almost twenty-five years before.[7] Another ship lay moored in the Bay of Whales; on its deck he could see aircraft displaying the unmistakable roundels of the Royal Australian Air Force.

He had been beaten to the Bay.

On 5 December 1935 the Prime Minister of Australia, Joseph Lyons, announced that a rescue attempt was being organized to locate the missing

Wilkins' Lockhead Vega (left) and Ellsworth's Northrop Gamma 2B (right) were featured in the popular Antarctic Territory series of the Australian Post Office.

aviators Lincoln Ellsworth and Herbert Hollick-Kenyon.[8] *Wyatt Earp* had received the last radio message from their aircraft, *Polar Star,* eight hours after take-off. The message itself was garbled but did not seem to indicate anything amiss with the flight; then it suddenly broke off and nothing more had been heard.

In a private message to the millionaire's wife, Wilkins had reassured Mrs. Ellsworth that the cessation of messages was no cause for alarm. There was nothing to indicate that the two aviators were in any peril. Moreover, their rescue plan was being followed. No doubt Wilkins feared that a furor in the press could cause matters to get out of control. He and Ellsworth had evolved a perfectly sound contingency plan. To have outsiders called in, or rather "buying in" to attempt a rescue, was the last thing either of them wanted. But the publicity had provoked considerable alarm; American newspapers editorialized on "what was being done." *Time* went so far as to suggest that Ellsworth may have switched off the radio himself to draw attention to his expedition. From Norway, a loyal Balchen vehemently rejected the magazine's assertion, while Admiral Byrd opined that the missing men were most probably safe and sound in Little America with a broken radio.[9]

What Wilkins had not anticipated was Australia's response to the emergency. Sir Douglas Mawson, the country's foremost Antarctic authority, and Captain John King Davis, Commonwealth Director of

Navigation and himself a veteran Antarctic mariner, told the government that a rescue plan needed to be organized at once. Australia was best placed to reach the Antarctic coast, but if they delayed the Ross Sea would be impassable after early February and a ship could not gain access, or could be iced-in.

Mawson led the initiative of the Australian rescue endeavors while his old skipper, Captain Davis, became officer-in-charge.[10] Their effort would be focused on Ellsworth's ultimate destination at the Little America site on the Bay of Whales. Next they needed a ship—but what ship? "Things are getting out of hand, could the British assist?" asked Davis in a cryptic memo that, supported by Mawson, went to the Prime Minister. The next link in the rescue chain reached across to London. From the Dominion Office to the Colonial Office in Whitehall, the question was put. The British government promptly offered the assistance of the Royal Research Society Ship, *Discovery II*.[11] On an icy edge beneath the Indian Ocean, far to Australia's south, *Discovery* lay off the coast of Enderby Land, pursuing oceanography and a whale count. The story goes that an officer came running along the deck as the scientists were about to lower the trawl net waving a message and calling: "Hey, just a moment . . . We're going to Melbourne. We have to find Lincoln Ellsworth."

But Australia's urgent interest in locating the missing pair, Britain's quick response with *Discovery II*, and New Zealand's pledge of cooperation were touched by more than the quality of mercy. All three governments were wary of "what the Americans are up to in Antarctica." (Byrd had been politely reminded through British diplomatic channels that his Little America was on New Zealand territory.) Speed and effectiveness in dispatching a rescue mission were seen as an all-important underwriting of the validity of British polar claims.

Wilkins, possibly unbeknown to Australia, had separately contacted the Discovery Committee asking that their ship in cruising along the

Above: When Antarctic expeditions sought public support in the 1930s, a fund-raising dinner brought together four men made famous by their polar exploits—(from left) Captain John King Davis, Sir Douglas Mawson, Vice Admiral Skelton, and Sir Hubert Wilkins.

coast listen for any wireless transmission from *Polar Star*. When he learned that Australia was taking a hand and intent on sailing *Discovery II* right into the Bay of Whales, Wilkins sent a signal to Davis expressing "sincere appreciation" for the organization but emphasized that, "unless unlikely delays are experienced, the Ellsworth Expedition's own adequate arrangements will meet all requirements." As a compromise he again suggested that *Discovery II* should stand off the pack, monitoring the ether for a possible distress call, while he would actually take *Wyatt Earp* into the Bay to effect whatever relief was needed. Despite his professed Australianism, Wilkins' first and foremost loyalty lay with the American who paid his salary—and who specifically wanted the trans-Antarctic mission accomplished in its entirety by his own private American expedition. Hence the detailed contingency plan.

As a sidelight to the rapid exchange of messages, on 7 December Prime Minister Lyons received a cable signed "Mary Louise Ellsworth" that said: "Let me express my personal gratitude to the government and people of Australia." Perhaps she did not share Wilkins' and her husband's confidence that the expedition could go it alone. At any rate, Australia was not for messing about. Not only would a ship be sent, but aircraft too. The Discovery Committee signed over their vessel to Australian management and by the time it docked in Melbourne on 13 December, after the 4,000-mile (6,400-km) voyage across the Southern Ocean, workmen were waiting on the wharf to begin the urgent task of clearing deck space so that two aircraft could be loaded.

The Australian government stirred into action. The Air Force told to produce two pilots and planes. A British ship and its crew ordered to Melbourne from the distant ice coast. What manner of man could prompt such rescue measures? I researched his life when writing *Moments of Terror* and found a debonair New Yorker, not given to heroics or self-promotion but absorbed with the spirit of adventure, be it the Wild West or a South American jungle, and whatever he did, money was not a problem.

Lincoln Ellsworth was born on 12 May 1880, into a family of Chicago wealth. His father, James Ellsworth, owned a string of coal mines which fed fuel to the thousands of steam locomotives on America's eastern railroads. Upon his father's death, Lincoln inherited a fortune and as a multimillionaire, he had no cause ever to work again. His, however, was a character of different fibre, tinged with a nostalgia for those tough cowboy hombres, like the gun-slinging marshal of Tombstone and Dodge City.

Having graduated in civil engineering from Columbia and Yale Universities, and possessing a strong interest in anthropology and archaeology, he found himself drawn to the remotest corners of the earth.

For a couple of years he took a job as an axeman in the survey party of a Canadian transcontinental railroad; as an engineer he then moved to distant mining camps. His wealth allowed him to raise a geological expedition into the Andes, reaching the headwaters of the Amazon River. Later in life he returned to South America, to search for the lost tombs of Inca emperors.

As with other polar leaders of his era, exploration of the Arctic posed the greatest challenge until he turned to Antarctica. (He had already attempted a North Pole flight with Amundsen and crashed; then flew over the Pole in an Italian dirigible.) Upon reaching Sydney at the close of his fourth expedition in 1939, he told a newspaper reporter:

> The Antarctic attracts me because of all the world it is the only place left where there is so much trailblazing still to be done—nearly a whole continent left where you can put out into the unknown, and land no man has seen before. When I was a boy I used to lie on the floor gazing at the blank spaces on the map. I was still a boy when I was taken to the memorial service for Scott in London. That was when I first decided I was going to be an Antarctic explorer. To my mind, there are three men in exploration—the trailblazer, the mapmaker, and the man who looks out for the resources of the country. I have been a trailblazer since boyhood and I always will be.

Ellsworth wanted his endeavours to be remembered for their value to exploration and science, not just as another stunt in this new age of the airplane. He had discovered and been able to report reliably upon a huge and hitherto unknown portion of Antarctica. He made position fixes of the features he found and fulfilled a cherished duty to fly the United States flag across the unknown ice. He reasoned that the highlands of the Antarctic Peninsula must be regarded as a continuation of the South American Andes, possibly linked with the Queen Maud Range and the mountains of Victoria Land that girded the polar plateau.

Allowing that this lofty chain formed a backbone across Antarctica, he questioned whether a sea level channel was likely to reach between the Weddell and the Ross seas. These were mysteries of Antarctica that he wanted his work to solve.

His 1935 flight was Antarctica's longest aerial journey until the Deep Freeze missions two decades later. No single flight had revealed so much of the seventh continent; it permitted a new map to be drawn by the American Geographical Society, proof of Ellsworth's dedication to the task. He frequently in times of peril repeated his favourite words from an old hymn: "So long Thy power has blest me, sure it still will lead me on." And from another verse, the final line that gave testimony to his restless nature: "Who has trodden stars seeks peace no more." Ellsworth died in New York on 26 May 1951.

The Royal Australian Air Force, now fourteen years old, had a select few aviators experienced in Antarctic flying through secondment to Mawson's BANZARE voyage of 1929–1931. Thus, when the government directed the Air Board to nominate a crew, the name of Flight Lieutenant G.E. (Eric) Douglas, one of Mawson's two pilots, was immediately submitted as leader. Davis supported the recommendation of Douglas as a "first class man"; one in whom the government could be sure of "not getting up to any stunts." Second pilot and navigator was the young Flying Officer Alister Murdoch; five other ranks were to be embarked for aircraft maintenance. The Board made two aircraft available—a de Havilland 60 Gipsy Moth biplane and a Westland Wapiti low-wing cabin monoplane. Both machines had floats and both were equipped with dual radio sets; the Moth was for short-range scouting, while the larger Wapiti had the task of dropping rations and a sledge to the missing aviators if the need arose. To save time it was arranged that the silver fuselages of the two aircraft would be painted yellow once the ship neared the Ross Sea.[12]

Crowds assembled on the dockside each day to watch preparation for the Antarctic voyage. Among the tons of stores and equipment, Mawson provided his wooden sledge, taken from the Adelaide Museum. Newspapers covered the loading of the two RAAF machines. Two days short of Christmas—and a day later than *Wyatt Earp*'s setting out on the other side of the world—*Discovery II*, under the command of Lieutenant L. C. Hill, RN, sailed down Port Phillip Bay to a send-off of cheers, tug whistles and blazing sunshine.

After a call to rebunker at Dunedin, the summer cruise quickly faded as they ploughed through the big seas of the Southern Ocean and by the second week of January had met with the Ross Sea pack-ice. *Discovery II*, a successor to Scott's famous *Discovery* of 1901, was an ice-strengthened vessel, but not an icebreaker. At times the crew had to line the rails, pushing out with long poles to deflect drifting floes from the vulnerable rudder and propeller as their ship went astern to prepare for another lunge at the white desert surrounding it. On 12 January 1936, Douglas and Murdoch took off in the Moth to search for a lead through the pack. Two days later the ship broke into the open waters of the Ross Sea. Douglas recorded in his diary:

> We are now 74°S and should arrive at the Bay of Whales sometime tomorrow and shortly after this the mystery may be cleared up. If Ellsworth is not at the Bay of Whales we will then endeavour to carry out some flying with the Wapiti . . . the ice blink from the Barrier is now visible around the horizon ahead. Air temperature down to 21°F (–7°C).

They were twenty-four days out from Melbourne when, from the crow's nest, the watch called that the cliffs of the Great Ice Barrier were in view. Fortunately the Bay of Whales was open, allowing Lieutenant Hill to sail them into the harbour on the morning of 15 January. Shouts went up as the crew pointed to machines of some kind parked close by the cliff edge; through binoculars they proved to be the tractors left behind by Byrd in 1934. Next, back from the cliff they spied the dark triangle of a small tent with orange markers fluttering beside it; Ellsworth was known to be carrying orange coloured signal strips. *Discovery*'s hooter sounded and signal rockets went up. Nothing stirred from within the tent.

Hurriedly the Moth was readied for take-off from open water beside the ship. Even on the short hop to Little America, Douglas and Murdoch suddenly found themselves enveloped in semi-whiteout conditions—"like flying in a bowl of cream," Douglas put it—unable to detect the surface until they saw another flag and then the tips of radio masts protruding through the snow.[13] They circled above the old campsite and saw a figure emerge from what appeared to be a trapdoor in the ice and wave his arms wildly at the plane. Murdoch dropped a small parachute attached to a bag of provisions and a note. "One of them is alive at any rate," Douglas reported on their return to *Discovery*.

"A Man Scrambled out . . . Started to Wave his Arms"
From the log of Wing Commander Douglas of 14 January 1936:

> At 8:30 P.M. the ship's officers reported they could see two orange-coloured flags and a tent some little distance in from the Barrier face . . . now as Ellsworth carried orange signal strips in his plane, it looked to me that they might possibly be living at Little America (situated about 7.2 km south over the Barrier ice) and erected this outpost as a signal to observers at sea.
>
> At 9:20 Al (Murdoch) and I were lowered over in the Moth and after a run of at least 800m I managed to get the machine in the air (air temperature 18° F) and then climbed steadily to 1,000 ft. I then nosed over carefully towards the Barrier face and set a course for Little America.
>
> As we progressed on over the ice the flying conditions became extremely bad and it was all I could do to keep the machine at a

A chateau in the Swiss Alps owned by the American millionaire Lincoln Ellsworth (right) provided a backdrop for the honeymooning Sir Hubert and Lady Suzanne Wilkins.

uniform height and on course. This was entirely due to the glare from the ice merging with the reflected glare from the clouds and we could not see the surface of the ice until we picked up another flag about 3.2 km in.

Shortly after this we saw what appeared to be a crevassed area, but as we approached it changed shape and we could then see that what appeared to be cracks in the ice were actually wireless masts and poles. As I circled this area we observed an orange-coloured ground strip placed near what appeared to be the top or roof of a hut. Then imagine our delight when a man scrambled out from this roof and started to wave his arms.

We continued to circle and after a few minutes I threw overboard a small bag of provisions attached to a parachute. This landed about sixty yards from the man who immediately walked across to it in his snow shoes . . . I turned east to look at an object that appeared to be a wing of an aeroplane. Sure enough it was the port wing of a monoplane and as I had heard that Admiral Byrd had taken home with him all his aircraft, I naturally came to the conclusion that this was Ellsworth's machine. I then headed away to the arc of water sky and in a few minutes could faintly pick out our ship. We cruised along the Barrier face observing this and the sea for a suitable place for men to walk over to Little America.

A scattering of snow petrels has the bird watchers scrambling. White body and white wings against the whiteness of a floe. Hard to track—home on the little black eyes and short black beak: one moment you see them and swish-swish, the next they've disappeared. So small and fragile they look against the overpowering ice. Reminds me of how small and fragile Ellsworth's aircraft must have seemed in the vastness of Antarctica.

Then killer whales appear along the edge of the Barrier. Cruel heads rise; dorsal fins knife the water; glistening black shapes effortlessly plunge. We watch for the reappearance; Olympics medals would be no problem for these. Suddenly a killer arches across our bows, one of our number jumps back with a yell, "that was close!" Memories, maybe, of Admiral Byrd's near thing in the Bay of Whales. *Kapitan Khlebnikov* forges ahead, shoving aside the drifting floes; a basking

Weddell seal reluctantly begins to wriggle towards the edge. The decision is between our bows and the lurking killers. "A difficult choice," agree the experts with cameras and binoculars leveled from *Khlebnikov*'s bridge. Ellsworth would also face a difficult choice; to turn left or right? Life could depend on it.

With skis and wing tips severely damaged, Polar Star *is gingerly hoisted aboard* Wyatt Earp *from a disintegrating floe in the Bay of Whales on 13 January 1934. Two years of frustration and setback will pass before Ellsworth's expedition reaches the Bay again.*

Sold to the Australian Government after its five years of serving Lincoln Ellsworth, Wyatt Earp *was refurbished to go South once again for the Australian National Antarctic Research Expedition. However, the old ship proved under-powered, rough, and uncomfortable, and the 1948 voyage to George V Land (pictured) was its last time in Antarctica.*

CHAPTER FIVE

The Rescue

After his fifth frustrated attempt to fly to the Bay of Whales, millionaire Lincoln Ellsworth turned to a quietly spoken English aviator to win him a place in the explorers' hall of fame.

Herbert Hollick-Kenyon ("Kenyon") was thirty-eight and married with two children when Sir Hubert Wilkins recommended him as a successor to Bernt Balchen, who had severed his connection with the Ellsworth expedition. Though London-born, Kenyon in World War I served in France with the Canadian Expeditionary Forces; being wounded twice did not deter him from joining the Royal Flying Corps, where he gained his pilot's wings.[1] He is recalled as the quintessential casual Englishman—neatly dressed, pipe smoking, sparse of conversation. In Canada's far north he had achieved 6,000 flying hours when he agreed to sign-on for Antarctica.

If Kenyon was not much given to small talk, three hours after the takeoff from Dundee Island he was sufficiently enthusiastic to send Ellsworth a note in the cockpit of *Polar Star*—"So this is Antarctica! How do you like it?" For himself, Kenyon was not much interested in personal fame. Only the day before, when adverse weather caused him to turn their flight back for a second time, he told an angry Ellsworth, "I understand that you would prefer Lymburner (the alternative pilot) on the next flight. That's quite all right with me." Ellsworth was not a man to act irrationally, and in Kenyon he recognized the experienced hand he might need when the inevitable emergency arose in flying across Antarctica. And so, on 22 November, they were on their way again in the trim little silver and orange aircraft, heading towards Ellsworth's first objective of 80°W.[2]

Climbing to 10,000 feet (3,000 m) and with an outside temperature of –30°C (–22°F) they swung away at Hearst Land from the spine of the Peninsula and, on a steady southeasterly course, began the crossing of the unknown continent. A line of mountains rose up that Ellsworth photographed and named the Eternity Range. Hollick-Kenyon noted that at times he was flying "hands off" in supremely fine weather conditions.

Eight hours out of Dundee Island, at an estimated 950 miles (1,530 km) from Little America, the radio—which unknown to them had been garbling their transmissions to Lenz, the wireless operator on *Wyatt Earp*—suddenly went dead. Hollick-Kenyon's note read, "Transmitter out of action. What shall we do?" True to his resolve, Ellsworth replied, "Keep on to 80."

Another mountain chain reared out of the plateau, shorter and more compact than the Eternities they had left behind, but possibly higher. Ellsworth named the most prominent peak, which he reckoned to be about 4,000 meters (13,000 feet), Mount Mary Louise Ulmer in honour of his wife. Sentinel Range he decided as the appropriate title for this giant stockade that stood guard over the silent white continent that extended on every side. The sighting of the new range meant that, in flying time, Ellsworth was passing his goal of 80 degrees longitude, beyond which the unclaimed portion of Antarctica began. Before that magic line he did not wish to set foot on the snow, not even to attempt a repair of the faulty radio. But fourteen hours after take-off, with a deepening haze spreading over the horizon, he made a signal to his pilot: the time had come for the explorer to leave his mark on this untouched crust of the ice continent.

Lacking a horizon or any feature on which to judge altitude, Hollick-Kenyon throttled back and with *Polar Star*'s generous flaps extended they waited for the bump that would signal they were down. Even with the pilot's expert care, *Polar Star*'s skis jarred so heavily against the smooth, hard surface that Ellsworth felt "my teeth would go through my head." They clambered stiffly out of the cockpit to find the fuselage "crumpled" from the landing. This, however, did not prevent further flying. Their position was fixed at 79°12' South and 104°10' West. Time had come to run up the Stars and Stripes and make a first territorial claim on behalf of the United States. The flag Ellsworth flew had been sewn by his niece Clare Prentice and presented to him before leaving New York. His description continued:

> We stood in the heart of the only unclaimed land in the Antarctic—in the whole world. I felt a very meek and reverent person. To think that I of all those who had dreamed this dream, should be permitted its realization! For the moment I lost all sense of the troubled beginning, had no thought for the journey ahead. I was content,

> grateful . . . So here I raised the American flag. The area extending from Long 80°W to 120°W I named James W. Ellsworth Land after my father. That part of the plateau above 6,000 feet I called Hollick-Kenyon Plateau.

They dubbed it Camp I, or Desolation, at an altitude of 6,300 feet (1,920 m) and an estimated distance of 650 miles (1,040 km) short of the Bay of Whales—"almost the most inaccessible point of the whole route."[3] Hollick-Kenyon cooked bacon and oatmeal on the primus stove while Ellsworth went for a stroll, but soon felt so overcome—intimidated, even—by the endless monotony of the plateau that he returned to the aircraft. They took sextant readings, tried to operate the radio and then with outside temperature around –25°C (–15°F) retired to sleep in the neat little weatherproof tent that had the silken sides sewn to a canvas floor. After nineteen hours they were refreshed and ready to go again. Warmed by Hollick-Kenyon's fire-pot, the Pratt and Whitney engine sprang back to life and from the granular hard-packed snow *Polar Star* lifted them into the air "in a swift, breath-taking lunge of fifty yards or so" at noon on 24 November. The sky still showed bright and clear and, apart from the radio failure that they could not solve, there seemed no reason that Little America should not await them within another five hours.

It was as if the Antarctic weather sensed the optimism of the intruding pair and decided it had been too lenient, by far. Within thirty minutes they were forced to land in the face of an oncoming storm. With *Polar Star* dug snugly into a snow pit, which was a further reason for choosing a low-wing machine, the tent at Camp II held them for three days, waiting for a clear sky. They spent the time trying to check their position with the sextant, which seemed to be working erratically, and cranking a hand generator in the hope that someone in the world outside would hear the weak signal of their trail radio and could tell loved ones that Ellsworth and Hollick-Kenyon were

The rare sight of two ships meeting at the Bay of Whales: Ellsworth's vessel, Wyatt Earp, *is moored alongside the ice, while* Discovery II *stands offshore.* Wyatt Earp *reached the Bay on 19 January 1936, five days after* Discovery's *arrival.*

still intact. They took off again on 27 November, to fly for a bare fifty minutes before failing visibility forced them down. No Little America today: for a whole week, a howling blizzard held them captive.

With Hollick-Kenyon's concurrence, Ellsworth had taken the hard-nosed decision to dispense with a meteorologist on the 1935 expedition, his reason being the impossibility of forecasting distant Antarctic weather.[4] They would land *Polar Star* to ride out the elements and, when the weather cleared, fly on again. His logic stood solidly enough—provided that a gale did not blow their tent away, that the fuel supply held out, and they were not lost or disabled. At Camp III most of these hazards appeared ready to visit them.[5]

For three days they remained in their sleeping bags as a matter of sheer survival from the intense cold and screaming wind which, but for iced-in anchor pegs, threatened to whip away their fragile tent with them inside it. They daily cooked two small meals and in the evening swigged from a bottle of grain alcohol Wilkins had slipped into their baggage. Fuel reserves were much lower than calculated; figures indicated the Wasp engine had been functioning at no more than two-thirds of its advertised speed and efficiency. Ellsworth began to lose the feeling in his left foot, due to a leather inner moccasin that contracted when moist, stopping the circulation. And he had to admit to Hollick-Kenyon that, because of the erratic performance of their sextant, he could not vouch for where they were. Nor, when it came to the time of leaving, would *Polar Star*'s motor start. This was their ultimate moment of terror.

Hollick-Kenyon was one of those silent men who contribute much yet leave a light footprint. He found the fault with Ellsworth's sextant in a loose index error adjustment screw and, having tightened it, worked out a compensating solution to restore accuracy. To get them off the ice, he connected the dead starter motor to the live batteries of their defunct radio and—presto!—the Wasp engine fired, ready to send *Polar Star* on the final 500 miles (800 km) to the Bay of Whales.[6]

Before they could leave, Ellsworth spent the whole of 4 December crouched within the aircraft, using bucket and pemmican mug to scoop out the hard-packed snow which drifted up every cranny of the fuselage. A storm returned next day and the excavation had to be faced all over again. Weather soon improved on the flight from Camp III. They passed from Ellsworth's own "Highland" to the continuing plateau of Marie Byrd Land (of Byrd's 1929 claimed territory) at which point their flight path reached above 80° South, the closest to the Pole that Ellsworth would ever come. Droning along above the unbroken ice sheet, they covered the penultimate leg in 175 minutes, noting an obvious downward sloping in surface elevation and the outbreak of huge crevasse fields, a sure indication of the approaching Ross Ice Shelf.

The landing at Camp IV was for a brief position fix, 79°15'S and 153°16'W. "What an afternoon!" Ellsworth wrote, "the snow sparkled like jewels. There was no wind. Once more it was good to be alive, for we were off the high plateau and on the Ross Barrier at last, 980 feet above sea level and only 125 miles from our destination, with fuel enough to reach it." His reference to fuel reserves proved a little optimistic. And the distance to the Barrier was more like 160 miles (256 km). Nevertheless, the final flight of sixty-five minutes brought them over the humped ice of Roosevelt Island, where a division of the flow of the Ross Shelf produces the Bay of Whales. Ellsworth's own description tells of the view that met their eyes when they gazed to the right:

> But at that moment something else was commanding our complete attention. All during this last hour of flight a great water sky had been building higher in the north. Then all at once, as we came past Roosevelt Island, we saw it—slate-coloured open water on the north horizon, looking almost black in contrast to the white expanse across which we gazed. The Ross Sea!
>
> There was a goal at which so much Antarctic exploration had aimed, and we had reached it. Behind this moment lay three years of planning work, and travel, heartbreak and hardship, failure, discouragement, and renewed determination—and at last there it was. We had crossed the continent from the Weddell Sea to the Ross Sea.
>
> At such moments in the story-books men are supposed to make memorable remarks, but in actual life behavior seems to be different.

> What happened was this: As soon as the open water appeared in the north, Hollick-Kenyon turned round and looked at me. I expected him to say something, but he did not. Nor could I think of anything to say. I stared back at him that was all. Then we resumed our individual tasks. After all, what was there to say?[7]

A few minutes later the engine spluttered and went dead. The eerie sound of rushing wind filled the cockpit as the pilot guided them towards the glittering white of the ice shelf. "Hollick-Kenyon picked his spot," said Ellsworth. "And at 10:03 A.M., local time, 5 December the *Polar Star*, like a weary bird, came gently to earth." They had covered 2,340 miles (3,765 km) since leaving Dundee Island a fortnight earlier and had flown across Antarctica for twenty hours and fifteen minutes.[8]

At first reckoning, the pair believed Little America II to be but a half day's walk away. But in what direction, amid this desolation of undulating ice? *Polar Star* was dug into the snow and a food cache piled on the wings. Rations posed a problem, too. Should they carry a minimum pack or drag a heavy sledge load of supplies? From the wing, Hollick-Kenyon espied what he thought to be a machine standing amid the snow and, beyond it, the outline of a camp; armed with pocket compass and sextant they set off on their snow shoes. This futile excursion set the pattern for the following ten days—the most nightmarish of the whole trans-Antarctic epic. The "machine" turned out to be one of Byrd's discarded oil drums and the "camp" but a jumble of distant pressure ridges. Day after day through twenty-four hours of sunlight and fog they kept searching for Little America,

pushing out from the stranded aircraft, stumbling back again, alternately dragging the laden sledge and abandoning it, utterly exhausted.

Until the moment that a strange rushing noise drew them towards a steep ridge. At the very edge they halted, staring through the mist to where another step could have easily taken them—to the Ross Sea breakers dashing against the ice far below. At least they had reached the Bay of Whales. But which way to turn to find Byrd's camp? Luckily they chose west.

Nothing could be seen of the buildings except for pipes, chimneys and masts protruding like dead men's fingers from the snow. They dug downwards beside a ventilator shaft, opened a hatch and scrambled into the quiet of Little America's radio shack, abandoned almost two years before. On 15 December, Ellsworth at last

Below: Blizzard-bound for a week on the polar plateau late in 1935, Ellsworth and Kenyon then faced the exhausting task of digging Polar Star *from beneath the snow. Yet Ellsworth gave thanks for choosing a low-wing metal aircraft, which they had been able to sink into the surface, to save it from being blown away.*

could say that his trans-Antarctic crossing was complete. He dug into his rucksack and produced a small flask of Napoleon brandy, his wife's gift for celebrating the moment of triumph. He and Hollick-Kenyon shared the contents—"the best brandy I ever tasted," wrote Ellsworth. "Brown, fiery, yet smooth as velvet. Hollick-Kenyon took a sip and really smiled." They had tramped over 100 miles (160 km) to find Little America II. *Polar Star,* in fact, lay only 16 miles (25 km) away, though in a direct line that was barred with crevasses.

Ellsworth and Hollick-Kenyon were never buddies. How they survived each other's company through the trials and tensions of those weary lost weeks stood as testimony to each man's self-control and to the respect they held for one another. In the hut sixteen feet (5 m) below the surface, Hollick-Kenyon continued his fastidious dressing. Each day he shaved and bathed himself in melted snow water, somehow he kept his trousers and jacket looking as if they had just emerged from a bespoke wardrobe. On one occasion, irritated by Hollick-Kenyon's stolid silence, Ellsworth complained, "don't you ever talk?" Replied his pilot: "I have a bad temper. I prefer not to."

Ellsworth had supreme confidence that Wilkins would adhere to their rescue plan. Apart from tramping to the Barrier edge, setting up the tent with orange markers, and scanning the sea, they could do nothing much else but wait. Radio sets had been stripped from the hut. Hollick-Kenyon kept the coal stove alight and foraged for rations, of which he found plenty. Ellsworth lay in his bunk, wondering about his numb left foot, feeling feverish spasms of hot and chill and quietly fuming over his reading glasses, left in the aircraft cockpit. Hollick-Kenyon stretched himself on another bunk, absorbed in one of the library of detective magazines they had found on the shelves, contentedly sucking at his long pipe and dredging candy from a tin beside him—"gurgle, swipe and crunch," as Ellsworth put it. Perhaps one of the bitterest moments between the two aviators arrived when Kenyon

Above: Lincoln Ellsworth (left) with Herbert Hollick-Kenyon, before the take-off from Dundee Island. Ellsworth described Kenyon's control of their aircraft as "miraculous." Right: The Royal Research Ship Discovery II, *as depicted on an Australian Antarctic Territory stamp.*

observed that his millionaire leader had used three matches to light his pipe, then drily commented, "You must be the president of a match factory." To which Ellsworth retorted, "You use a good many more matches than I do."

So they existed through one week, and another, and another. Ellsworth was asleep when the sound of an aircraft brought Hollick-Kenyon scrambling from the hatchway on the morning of 14 January. To his surprise, the machine circling overhead displayed not the more likely American markings but, to him, the familiar roundels of a British military plane. The note in the container of food that fluttered down on a small parachute advised that a boat party would come ashore and if able, he should go to meet it.

Crew members of *Discovery II* confessed to being somewhat disappointed when to their cheers and clapping, Herbert Hollick-Kenyon came alongside in the ship's motor launch. They had expected a bearded, emaciated and bedraggled explorer. Instead here was a beaming, ruddy-faced, clean-shaven and check-shirted Englishman who stepped on the deck—as a figure the nearest one could imagine to a well-dressed polar tailor's dummy.

"Well, well! The *Discovery*, eh," Hollick-Kenyon exclaimed, accepting the offer of a whisky and soda. "This is an affair! But I say, it's awfully decent of you fellows to drop in on us like this."

His rescuers were so taken aback at this bizarre interlude that the conversation seemed about to lapse until someone asked, "Where's your aeroplane?" Hollick-Kenyon settled himself in a chair, deliberately filled his pipe and replied:

"The aeroplane? Oh, that's twenty miles away on the Barrier. You see, we ran out of gas."

"Did you crash?"

"Good Lord, No! We landed perfectly safely and walked in. The aeroplane's all right. We'll have to go and get it presently . . . Food?

Members of the Australian rescue party assemble around the hatch that led to the hut where Ellsworth and Kenyon sheltered for thirty-one days. The radio masts of Byrd's original Little America were gradually disappearing beneath the snow.

Oh, rather. Any amount at Little America. So much we hardly knew what to do with it. No, we weren't worrying. We knew *Wyatt Earp* would come along sooner or later . . . Pity about the radio. The transmitter switch went wrong . . . Yes, we landed all over the countryside during the flight. Whenever the weather got a bit thick you know . . . Ellsworth? Oh, he's all right. He's got a bad foot but he'll be along tomorrow."

'So you had no hitches at all?"

"No, none at all. Oh, yes! One slight hitch. Ellsworth left his spectacles in the aeroplane twenty miles away so he can't read, poor old chap!"

Six men of the launch party, led by Hollick-Kenyon, returned to the Barrier and tramped across the icy path to Little America to meet the missing millionaire. Ellsworth preferred to spend the night hours resting before making the journey to *Discovery*. They found the man to have a temperature of 104°F and his blistered left foot was in danger of turning septic when they carried him to the ship's sick bay next morning. On the same day, the signal went out on *Discovery*'s radio, to be picked up by the world press—"Ellsworth Found!"[9]

Above: Ellsworth and his Australian rescuers stand on the snow crust above Little America base. Ellsworth suffered an infected left foot and had to be carried on a sledge, yet claimed, "at no time were we lost."

Right: To thank the Commonwealth Government for finding him, Ellsworth came to Australia aboard Discovery II. *He stands next to Captain L.C. Hill, with other members of the RAAF team around him. Pilot Eric Douglas is on Ellsworth's left, and Alister Murdoch stands beside Captain Hill.*

DISCOVERY II
PORT STANLEY

Eric Douglas described the American as a man of slight build, face burnt brown by the Antarctic sun, quiet mannered and modest of speech. Except that Ellsworth did not want to regard the Australian airmen and the British crew as his rescuers. Despite his sincere appreciation of the doctor's treatment and the warm bed given over to him in the chief scientist's cabin (the doctor also had found a pair of spectacles that he could use), Ellsworth politely insisted that the whole Australian effort was really unnecessary; his own relief operation, true to its schedule, was close at hand.

On 19 January Ellsworth's faith in Wilkins' management was borne out when a small wooden ship flying the Stars and Stripes nosed around the headland into the Bay of Whales. Indeed after a 5,500-mile (8,800-km) voyage from the Antarctic Peninsula, probing the ice at various prearranged places to check for Ellsworth's presence, Wilkins had reached his leader three days ahead of schedule. A moment of humour came when Ellsworth chided Wilkins for flying *Wyatt Earp*'s flag at half-mast with the comment that the report of his demise was somewhat exaggerated. "You flatter yourself," Wilkins replied. "We heard on the radio that King George V has passed away."

Wilkins came aboard *Discovery II* and sat next to Douglas at tea; their conversation led to another diary entry by the Australian airman:

> I found him a charming man though I have no doubt he is a bit of a stunt merchant. He was a bit annoyed with the relief expedition (ours) and his reasoning appeared to be quite OK until I found that he had wired Washington asking the British Government to help with the *Discovery II* by this ship steaming south and listening for W/T signals. He could hardly blame the Discovery Committee and the Australian Government for wanting to carry it out properly or not at all.[10]

The crew revived one of Byrd's abandoned tractors and loaded it with gasoline for the journey to where *Polar Star* lay buried. They dug out the aircraft and Hollick-Kenyon flew the remaining distance to land on a floe beside Ellsworth's ship. Now it could be said the trans-Antarctic feat was really and truly accomplished. From the cockpit, Wyatt Earp's gun belt was restored to its owner, who drily observed that *Polar Star* rather like himself had journeyed 65,000 miles (104,000 km) by sea to spend but twenty hours making history in the sky. The machine, looking so small and fragile, was loaded on *Wyatt Earp*'s deck for the voyage home, eventually to find a place of honour in the Smithsonian Museum.

The RAAF team regretted that Hollick-Kenyon had to accompany *Polar Star*. "We are very sorry he cannot come," wrote Douglas, "he is such a fine chap, very modest and cheery." Douglas also presented Ellsworth with the small parachute they had dropped on Little America. Sergeant Easterbrook, one of the RAAF fitters, inscribed the fabric with the words "Compliments to Messrs Ellsworth and Kenyon from the personnel of the Royal Australian Air Force aboard the RRS *Discovery II*, Bay of Whales, Ross Sea, Antarctica Jan. 15th 1936. Antarctic Parachute Co. Never-Failed-Yet." At the bottom of the message, all the RAAF team added their signatures.[11]

Mawson and Davis were waiting to welcome the world-famous American millionaire when the ship reached Melbourne on 16 February. Ellsworth was interviewed, photographed, feted, taken to Mount Buffalo resort and then flown to Canberra for lunch with Prime Minister Lyons and his cabinet.[12] He joined the liner *Mariposa* in Sydney; at Honolulu his wife awaited him for the remainder of the voyage to Los Angeles where an honour guard of aircraft came out to escort the ship. At New York in April, a celebrity's welcome greeted both Ellsworth and Hollick-Kenyon when *Wyatt Earp* docked in Manhattan with *Polar Star* proudly displayed on deck.[13] On 16 June 1936, the United States Congress voted Lincoln Ellsworth a special gold medal—"for claiming on behalf of the United States approximately three hundred and fifty thousand square miles of land in Antarctica between the eightieth and one hundred and twentieth meridians west of Greenwich, representing the last unclaimed territory in the world."

Matters were rather more mundane in Australia where a totalling of rescue expenses showed that the RAAF effort had cost the taxpayer £2,696.4.7d, and after allowing for contributions to *Discovery II*'s expenses, an amount of £134.9.8d was owed to the British government.

Once as a young newspaperman in Sydney I had been sent to Darling Harbour to write of a small and unspectacular wooden vessel bearing the name *Natone*. She was engaged in the New Zealand potato trade and was lost only a few years later off the Queensland coast.

"Natone" did not seem to mean much, but if one reached back into maritime history, *Wyatt Earp* meant a great deal. Under her former name, the little ship that began life in the Norwegian herring trade had carried Lincoln Ellsworth to Antarctica. Twice she had sailed to the Bay of Whales, her deck laden with the sleek low-wing monoplane that the New York millionaire had chosen for his trans-Antarctic flight. Among those who had occupied her cabins were Sir Hubert Wilkins, the indomitable Australian polar explorer, and Norwegian-born Bernt Balchen, first man to fly an aircraft to the South Pole; in New York harbour she had returned to a hero's welcome. Still visible on the side of her smokestack was the little trapdoor into which they shoveled ice and snow for the melt water tank. Ellsworth chose the name *Wyatt Earp* and on the bulkhead of the owner's cabin hung the gunbelt of the Arizona Marshal who was his Wild West hero. In 1939, at the end of Antarctic adventuring, Ellsworth sold his ship for £4,400 to the Royal Australian Navy, which, under the new name of *HMAS Wongala*, began wartime patrolling of Gulf St. Vincent in South Australia. For another two years she was a Sea Scout training vessel stationed at Port Adelaide. In February 1948, once more appropriately renamed *Wyatt Earp*, she carried Australia's first postwar expedition to the Antarctic coast. In the Navy dockyard she had been refitted with a Crossley 450 horsepower diesel engine and accommodation was enlarged to house thirty men. Cosmic ray measuring equipment plus magnetic observation recorders and radar were installed, and deck space found for a RAAF Vought Sikorsky Kingfisher aircraft. But the wooden timbers that once defied the Ross Sea ice were aging, her cargo space insufficient, her engine power inadequate, and at the voyage's end she was sold to a commercial operator. When lost near Double Island Point in 1959, a large slice of Antarctic history was interred with her bones.[14]

Top: President Roosevelt presented Ellsworth on 16 June 1936 with a Congressional gold medal as recognition for his Antarctic exploits. Above: Natone *(the former* Wyatt Earp*) as last seen at Darling Harbour, Sydney.*

Artist Ken Gilroy's impression of the Bay of Whales in 1947, taken from a Navy aerial photograph with three Highjump vessels moored at the southern ice edge.

We are reminded that Ellsworth and Wilkins returned to Antarctica in December 1938 with a bigger aircraft—Northrop's latest Delta cabin monoplane, and a new pilot, Jack Lymburner, who had been Hollick-Kenyon's deputy (Kenyon had joined Canadian Airways), but with the same old ship. This time, at the age of fifty-eight, Ellsworth's objective was a super-transcontinental flight from the vicinity of Kemp Land in Australian Antarctic Territory to the South Pole, and then returning to the Bay of Whales where Wilkins would once again meet him with *Wyatt Earp*.

However, a belt of thick pack ice extending for some 840 miles (1380 km) below the Indian Ocean seriously delayed the ship, forcing an abandonment of the aviation program—except for a single interior flight. On 11 January 1939, Ellsworth and Lymburner reached 70°S into Princess Elizabeth Land, whereupon Ellsworth dropped a copper canister claiming all the ice beneath them—about 380,000 square miles, according to Ellsworth—for the United States Government, and bestowing on it the name of American Highland. "Not a very nice way for him to show his gratitude to Australia," says one of our company upon hearing the story. But we are afforded some consolation in learning that Wilkins quietly deposited a paper reaffirming Australia's sovereignty over its Antarctic Territory each time he went ashore.

Tales such as these are traded aboard *Kapitan Khlebnikov* around coffee in the library, or in the lecture theatre where travellers are treated daily to expert commentaries on polar history, penguins, whales, seals, volcanoes and glaciers, which are all around us while we sail the Ross Sea. But lectures today have to take a second place! We have met with B15A, a major portion of what is reputedly the world's largest observed iceberg—a slab of ice so vast, in fact, that its existence could only be detected and tracked from satellite images.

DECEMBER 31, 1956
PACIFIC EDITION
PROSPERITY and EXPANSION
A Business Review and Forecast
TIME
THE WEEKLY NEWSMAGAZINE
ANTARCTIC EXPLORER SIPLE
AUSTRALIA 2/6
BURMA 1 kyat 50 pyas
CEYLON Re. 1.50
FIJI ISLANDS 2/3
FORMOSA NT$ 7.00
HONG KONG HK $2.00
INDIA Re. 1/8
INDONESIA Rp. 6
JAPAN 100 Yen
KOREA 150 hwan
MALAYA ST. $1.00
NEW ZEALAND 2/-
OKINAWA RY 30
PAKISTAN Re. 1/8
PHILIPPINES 80 centavos
THAILAND 7 Baht
U. S. ARMED FORCES 25 cents
VIET NAM 10 piastres

CHAPTER SIX

Boy Scout Bay

No one knew the Bay of Whales better than Paul Siple. In a memorable report, he accurately forecast an event of which the world might stand in awe: "Within four or five years the navigable portion of the Bay will be completely compressed and by that time a great cataclysm will be close at hand."

When this tall, big boned, and friendly man took up his post as Science Attaché at the American Embassy in Canberra, few people, apart from the small "Antarctic community," were aware of his background. Dr. Paul Siple was one of the world's foremost living polar explorers. A few years before

Left: Paul Siple's polar fame put him on the front cover of Time *magazine, 31 December 1956. He was first leader of America's station at the South Pole.*

Above: McMurdo camp at the establishment of Operation Deep Freeze in 1957–1958. Below: Dormitory blocks of the base that now houses some 1,000 Americans each summer season.

coming to Canberra, he had been scientist-leader of America's first station at the South Pole. Going back some twenty years he had commanded West Base at the Bay of Whales. And going back ten years before that, you would have found him as a young Eagle Scout, skinning seals and penguins at the first Little America.

I would visit the American Embassy to sit across the desk from the man whose face was once on the cover of *Time* magazine, to prompt him—for he was ever modest and some "prompting" it took—to recall moments in his thirty-year association with Admiral Byrd and the Bay of Whales.

Out of all the senior Boy Scouts in the United States, he had been chosen in a competition, he said, aged nineteen, to join the Byrd Expedition.[1] He remembered the wild voyage to Antarctica in *City of New York* and wondering if he would be allowed to winter-over at the base. Fortunately his tall and muscular build persuaded Byrd to appoint him as the expedition's "amateur taxidermist" and he spent much of the year cutting up seal and penguin remains for study in American laboratories.

On the same 1929 expedition he was one of Byrd's party in the motor boat when they were pursued by a pod of killer whales. He still smiled at the memory of Byrd pulling out his revolver when they took refuge on the ice edge—"to one of those killers, a bullet would have hurt as much as a mosquito bite." The young "taxidermist" was not allowed to forget the incident. "Every time I went into the mess hut, someone would yell, "When are you going back to the Bay to slice up one of them killers?""

Above: The Eagle Scout salutes his leader and Dr. Siple (right) dressed for the weather in later years.

Though Siple's next Antarctic posting designated him chief biologist at Little America

II (he had graduated in biology and geology since Little America I), it was his skills at navigation and dog handling learned on the 1929 expedition that won him leadership of a four-man sledging party making a seventy-seven-day journey into unexplored Marie Byrd Land.[2]

> There was something about the Antarctic that I found exhilarating on the trail. Its quiet was so profound that one could spend hours on end in satisfying contemplation. At the age of twenty-four, here I was getting my first real taste of traversing into virgin unmapped territory . . . a form of exploration not far removed from that of the early men who sought to achieve the Pole. . . .

During the second expedition, he was also a member of the team that drove 120 miles (192 km) across the Ross Ice Shelf in blizzardly winter darkness to rescue Admiral Byrd from his one-man observation post known at the Bolling Advance Weather Base. They waited for two months before Byrd, who had been slowly poisoned by carbon monoxide fumes from his stove and generator, was well enough to make the return journey in the Citroen tracked vehicles to Little America II.

Siple regarded the expedition of 1934 as tribute to his leader's drive and determination against huge odds.[3] "Byrd drew upon a wave of popular support and generous donations when we set sail the first time for the Bay of Whales. But by the time of planning for Little America II, the Depression had arrived. Yet he still made it a reality."

Once again Edsel Ford was a backer, as was John D. Rockefeller, Jr., and Joseph Pulitzer, publisher of the St. Louis *Post-Dispatch*. To their names would be added those of William Horlick, the malted-milk king; C.R. Walgreen, head of a big drug company; and Jacob Ruppert, a brewer and owner of the New York Yankees; along with the corporate support of General Motors, International Business Machines (IBM), American Airways and many others from the "Who's Who" of American industry. Ralston Purina Cereals donated fifty tons of dog food for the huskies.

With best wishes
to another venturer
to Antarctica David Burke

Paul A. Siple

9 Sept '61
Sydney

A memento of knowing Paul Siple.

With aerial conquest of the South Pole behind him, Byrd set out to explore in more detail what he termed "the Pacific quadrant," the vast, unchartered and unclaimed region lying between the 80th and 150th West meridians where his 1929 expedition had only picked at the edges.[4]

Byrd's second expedition consisted of two ships: *Bear of Oakland* and *Jacob Ruppert*. To carry his aerial survey parties he brought four aircraft, supported by five tractors, two of them developed originally for desert travel by the French automotive genius Andre Citroen. Foremost among his airplanes was Curtis Wright's new Condor, a large twin-engine biplane given the name *William Horlick*. In appearance the Condor looked slow and cumbersome, but in operation it proved extremely useful, being adaptable to either skis or floats and having enough power to lift a 19,000-pound gross load. The other aircraft were a high-wing Fokker, a Fairchild and a Kellet K-4 autogiro, which was the first rotary-wing machine to operate in Antarctica.[5]

The expedition, numbering in the wintering-over party 56 men, 153 dogs, 3 cows, and a calf (born during the voyage south), went ashore at the Bay of Whales in December 1933. Siple recalled the back-breaking work in digging out Little America I before they could reoccupy Byrd's old base and new buildings erected to connect with those that were half-buried, thus bringing into existence Little America II. Among the base amenities was a theatre "wired for sound."[6]

Byrd was a master of public relations. To bring Antarctica into the homes of the American people, or to be specific, the amazing and heroic deeds of the Admiral and his expedition, he began the first direct-voice

On his third Antarctic expidition, Paul Siple was leader of West Base—Little America III—at the Bay of Whales. One of his exploring parties in 1940 surveyed the Saunders Mountains on the Pacific coast region of Marie Byrd Land.

broadcast from Little America II to listeners in the United States on 1 February 1934. Weekly programs from "the ice" were then featured on the Columbia Broadcasting System's network, while newspapers received almost daily bulletins by radio.

The second expedition placed greater emphasis on science and surveying. Was Antarctica in reality two continents? Did a strait extend from the Ross Sea to the Weddell Sea? Did the mountains of the Antarctic Peninsula link up with the Queen Maud Range? Siple frequently filled the role of navigation expert. Long-distance flights were planned to unravel these mysteries. With Harold June as chief pilot, the Condor covered 770 miles (1,240 km) on a mission to define the eastern edge of the Ross Ice Shelf; in the process, Byrd named the William Horlick Range. Another flight of 450 miles (720 km) to the north-east penetrated the immense ice dome which he had called Marie Byrd Land, in honour of his wife. Territorial claims to the region were volunteered on behalf of the U.S. government. The surveys covered some 450,000 square miles (1,165,000 km) of Antarctica, of which more than half had never been previously sighted.

Aerial photography was utilised through having a special camera bay built into the Condor's cabin. Byrd's crew refined the technique of dropping smoke bombs to detect surface winds and the elevation of the ice before attempting a landing. Seismic soundings were carried out to determine the depth of the ice crust; the improvement in radio technology enabled his men to communicate freely among base, aircraft, and field parties. The Kellett autogyro, which had made its first flight on 1 February 1934, proved an extremely useful workhorse until lost when it spun into the ice eight months later. The

Fokker also crashed during an attempted take-off. Men were injured, but Byrd's remarkable record endured—no fatalities.

After returning home early in 1935, Byrd had worked hard to lift the awareness of America and its government towards a new zone of national interest. Approaching 1939 he planned his third private expedition—to "fill in some of the blanks in the map," as he put it.[7] However, times were changing. The world was a much more uncertain place than in the Indian summer of the 1920s and even in the slumped Depression years at the turn of the decade. Big Norwegian whaling fleets were on the polar seas, reaping a rich harvest and using light aircraft to name the coast on which they had already raised King Haakon's flag. What might the French do next? In 1924 they had claimed Adelie Land, directly to the south of Australia. At Adolf Hitler's orders, two Luft Hansa flying boats were making for the ice.

Alerted to these events, particularly because the Germans were intent on exploring Antarctica, President Roosevelt decided to establish the United States Antarctic Service around which a government-sponsored expedition would be organized. Congress authorized a budget of $350,000 to underwrite the venture, which signified the end of America's political "disinterest" in the frozen South. Byrd's ten-year investment in public opinion had paid a dividend. Quickly he threw in his lot with the government initiative and on 30 June 1939 was appointed commander of the first officially sponsored American Antarctic expedition since Commander Wilkes of the U.S. Navy had sailed south a century before.

One day the phone rang in Siple's office. Byrd's voice was on the other end and as usual he lost no time in coming to the point. "Paul, there's a new expedition going South. We want you to come." The Antarctic Service expedition of 1940 consisted of two ships, *North Star* and *Bear,* to carry the 59 men, 160 dogs, and 8 aircraft for wintering-over. Government finance permitted two establishments—once again a Little America or "West Base" on the Bay of Whales, commanded by

Unloading Dr. Thomas Coulter's mammoth Snow Cruiser from North Star *in 1940. Designed as a mobile base, the ungainly thirty-five-ton machine could barely climb the slope from the Bay of Whales to Little America III—and went no further.*

Paul Siple, and located five miles (8 km) north of the previous Little America positions, and a second outpost known as East Base on the Antarctic Peninsula.[8]

Another significant difference allowed under government auspices was the declaration of America's territorial desires upon Antarctica. They focused specifically on the vast quadrant of Marie Byrd Land, which, to one side was bordered beyond 80° West by Graham Land as the British called it, and to the other beyond 150° West by the Ross Dependency, administered by New Zealand. Roosevelt specifically instructed Byrd to show the American flag at a series of points during exploration sorties. In the words of the presidential instruction: "so that in the event of Congressional action, the region explored by the United States may be claimed as a national territorial area."[9] In a vindication of Byrd's previous work, the expedition's purpose was also "to give America full and up to date knowledge of Antarctica, especially the hitherto unknown sector [beneath] the extreme southwest part of the Pacific Ocean. It is incumbent upon us to be prepared with information for whatever policy concerning territorial rights the government may decide upon."

Noteworthy, too, was Byrd's response to the president's instruction. On his 1935 return from Little America II he referred to Antarctica in the context of "let there be no boundaries here. Let the Antarctic stand as a symbol of peace and a beacon for the world." Subsequently he was to modify that view and Antarctica became potentially "a base for commercial and naval aircraft, a key to weather forecasting, and a source of huge mineral deposits." Further, America should regard any attempt by another country to establish a base around the Pacific zone as "an unfriendly act."

At the Bay of Whales, Siple faced the challenge of making a new Little America III to the east of the original site, now covered almost to the tips of the tall radio towers erected in 1929. The attitude of a government expedition was much more inclined to the bureaucratic workaday, more serious and disciplined than in the era of the privateers. Likewise the list of sciences represented had expanded impressively over the intervening ten years, now embracing geology, glaciology, meteorology, geomagnetism, auroral observation, seismic and cosmic ray studies, biology and radio communications.

Three aircraft comprised the expedition's aviation wing at Little America—again, a heavy Curtis-Wright Condor biplane; a Barkley-Grow seaplane, used in three important flights during *Bear*'s cruise off the Marie Byrd coast in February 1940; and a small Beechcraft monoplane, intended to be carried on the roof of the Snow Cruiser, a massive over-snow vehicle which proved a total failure.[10] In the summer of 1940 the Condor completed two lengthy missions, one across a remote part of Marie Byrd Land and the next in a survey of the eastern flank of the Queen Maud Range. During the second flight the crew discovered another huge river of ice, the Shackleton Glacier, flowing from the plateau and were able to check on the estimated 15,000-foot (4,600 m) height of Mount Alton Wade, which Byrd had sighted in 1929 on his way to the Pole.

During one cosmic ray observation flight, the Beechcraft reached an altitude of 21,000 feet (6,400 m) a record for Antarctic aviation. One of the longest flights of 1940 brought back a photo-survey of 1,000 miles (1,600 km) of the Trans-Antarctic Mountains. Photo-surveying had reached high accuracy while aerial color photography was also introduced experimentally to Antarctica. Map definition was heightened through radio bearings and position reporting to base and, through radio communication, in greater coordination with surface parties.[11]

Though Roosevelt intended a permanent American presence in Antarctica, the third expedition closed up Little America III and East Base on the Peninsula in March 1941.[12] The Bay of Whales once again returned to its lonely silence. Another catastrophic world war had commenced. Before the year's end, the Japanese would attack Pearl Harbor and an America at battle filed away its potential involvement in the seventh continent for another day. Byrd returned twice more to Antarctica, but the days of the derring-do commander were finished. Future American expeditions were of such massive proportions that only a military machine could support and direct them.

Between his second and third Antarctic expeditions, Siple achieved his Doctorate in Geography, and had wed Ruth, his college girlfriend of six years past. Their marriage produced two daughters. In World War II he was an Army cold weather adviser, which led him to active duty at the battlefronts, gauging the health and well-being of soldiers fighting under low temperatures; afterwards his work, which continued with the Army, was similar in the Korean campaign. Polar experience provided the basis for his development of the "chill factor" scale, which sought to understand the impact of severe cold weather on the human body at different levels of temperature and wind.[13]

Though Admiral Byrd was twenty years his senior, the friendship between the two men did not fade—"we progressed from a father-pupil relationship to one of mature and enduring comradeship"—and it was Byrd who would again summon Siple when Antarctica called men South.

Paul Siple returned to the Bay of Whales in 1947 to advise America on fighting Arctic battles should the cold war turn to a shooting war. In a world barely emerged from bloody conflict, the

Right: When Admiral Byrd announced "Little America is the most peaceful spot in the world, due to the absence of women," he aroused the ire of feminine America. On his last visit to Antarctica in 1955, he was confronted by young women waiting for him at Dallas airport.

Byrd
UNFAIR
To American
BYRD

path to peace proved dark and gloomy. An iron curtain had descended across half of Europe, whole nations were silenced, and the arms race was far from dead. Against this forbidding backdrop, Operation Highjump descended upon Antarctica with all its 4,700 men, 13 Navy ships, 23 aircraft, and a host of other military machines. Admiral Byrd, Officer-in-Charge, pulled no punches when he described the reason for playing war games at the bottom of the world:[14]

> The object lesson of all this is obvious. The shortest distance between the new and old worlds is across the Arctic Ocean and the north polar regions. It is freely predicted that here will be one of the great battle areas of future wars.

Despite Admiral Byrd's grim forecast, made amid the antagonisms and suspicions of the cold war period, an interview of 1964 I had for the *Sun-Herald* with the leader of the Soviet Antarctic expedition seemed distinctly devoid of sabre rattling.

"Your health!" said the Russian and clinked his glass of wine against my own.

We drank the toast. "The Arctic was my first love," he said "Now it is the Antarctic. It is like another planet down there—it is like being on the moon."

Last week Sydney had a visit for a day from the man who directs Soviet Russia's program on the ice continent 2,000 miles south of Australia. For twenty years he was studying the Arctic oceans and made history by drifting for 376 days on an ice floe around the North Pole.

He led the first Soviet expedition to Antarctica and established Mirny, the expedition's headquarters and other bases that are scattered across Australian Antarctic Territory and spent 15 months living in them.

I met him waiting in the lounge of a leading city hotel—

Breaking into the Bay of Whales in the summer of 1947 to make way for the force of 4700 men, 13 ships, and 23 aircraft that comprised Operation Highjump.

Dr. Mikhail Somov, geographer, oceanographer, and one of the foremost present-day polar explorers. Somov is regarded by some as the "mystery figure" behind Soviet penetration of Australian Antarctica, from which come startling rumours of permanent Red settlement, a territorial "grab," and even wilder rumours of submarine bases and rocket sites.

Somov is tall and slim, and rather darkly good looking (women are said to find him decidedly attractive). When I met him he wore an American-style single breasted suit of dark blue fabric, and of his achievements he was certainly modest. As he relaxed in a deep lounge chair, pushing back iron grey hair and nodding readily in answer to my questions, it would have been hard for anyone to guess his powerful title—Deputy Director, Institute of the Arctic and Antarctic, Leningrad.

An Antarctic Weather Symposium in Melbourne had brought him to Australia for the first time, followed by a general meeting of SCAR (Special Committee on Antarctic Research) in Canberra. Those appearances concluded, he and his companion, Professor B. L. Dzerdzeyevsky, a bearded, bespectacled weather expert from Moscow University, wanted to spend the day being a couple of Sydney tourists before returning home next morning. (Ironically, while Canberra commentators were speculating about the Red flag flying in Antarctica—and how it would figure in resumed diplomatic relations between Australia and the U.S.S.R.—the man who planted it there was swimming in the surf at Newport Beach.)

I haven't an inkling what goes on at the back of Somov's mind, or what game the Kremlin may be playing in the ice continent, but the few hours of our city meeting couldn't have been more friendly or, so it seemed, more open. As far as I know, it was the first time at least in recent years that he had been interviewed outside Russia. Russian is his only western language, which was spoken in a pleasant, well-modulated voice, while Dr. Dzerdzeyevsky (pronounced jer-jay-evsky) grappled with English translations.

For a start, Somov went upstairs to his room to get an Antarctic map that would make our discussion easier. He spread it on the table between us, a glossy folding chart covered with lines, distances and Russian characters. "This is how much we have mapped ourselves," he said, pointing to a pattern of symbols that extended from Cape Adare (on the Ross Sea, south of New Zealand) to Mawson base, beneath the Indian Ocean. Between these two points, a mighty inverted red crescent underscored the entire coastline of

Australian Antarctic Territory. "We heard your Australian expedition last month made its first landing on the uncharted coast of Oates Land," said Somov. "We have already been there—I do not think they knew that."

None of my questions went without a response: the size of the Soviet expedition, the plan for exploration, the lowest temperatures encountered, the thickest ice . . . "We have ninety-three, maybe ninety-five men in the south. At Vostok on the plateau, the lowest temperature our men recorded was minus 125.3 Fahrenheit." At this point Dzerdzeyevsky interrupted with "it was so cold the petrol turned to jelly."

"Near the Pole of Inaccessibility," Somov continued, "we found the ice was 12,250 feet thick. That is a record, too, deeper even than the ice at the geographic South Pole." (I tried to imagine the State of New South Wales covered by a block of ice more than two miles high, with all of us living on top of it; that would be the equivalent to the Russians' finding.)

Their next major exploration was planned to strike 3,750 miles across the Antarctic plateau, using big tractors built in the city of Kharkov. "We have three Kharkovchanka at Mirny"—they pronounce it "Meerny"—"and probably all will be used," said Somov. "They have two drivers each, a navigator, a radio operator and can accommodate twelve scientists. The men live inside the Kharkovchanka, which contain bunks, a kitchen, armchairs, radios, a snow melter, and laboratory. Large sledges loaded with fuel and supplies are towed behind. Our machines are rugged, 35 tons weight and 550 horsepower to propel them.

"They will travel via the Geomagnetic Pole, the South Pole and the Pole of Inaccessibility. Surface elevation rises to 14,000 feet and substantial crevassing is anticipated." I commented that it appeared to be a risky journey with a high chance of things going seriously wrong. Dzerdzeyevsky interrupted with a shrug—"Everywhere one goes in Antarctica is an adventure and dangerous."

Somov smiled and turned to the lighter side. "At the South Pole we will look forward to celebrating our arrival with the Americans who are stationed there. We are all friends in Antarctica."

The opening to the Bay of Whales was barely 300 yards wide when Siple led an advance detachment of Operation Highjump to commence Little America IV, with the priority task of building a landing strip for the aircraft scheduled to make the first fly-in to the Antarctic continent. Some 750 miles (1,299 km) off the coast, *USS Philippine Sea* turned into the wind at 30 knots (35 mph or 56 kph) and prepared to launch its six R4D transports, the largest planes to attempt a take-off from the restricted space of a carrier deck, and the first equipped with combined wheel-ski landing gear. Admiral Byrd was aboard the first plane, which, at full throttle and with a fiery blast from its JATO rocket boosters, lifted into the air after a frighteningly short run of 400 feet (130 m). One by one the other aircraft followed, all of them some four hours later safely reaching the Bay of Whales, where Siple recalled that he had anxiously watched storm clouds blowing in from the south.

Byrd repeated the journey he had made eighteen years before when he joined Commander Trigger Hawkes in an R4D flight on 15 February to the South Pole—looking down at the white desert of the polar plateau he might have remarked that "nothing much has changed." However, things were changing for Byrd. The reality was that in the estimation of many in the present-day defense hierarchy, the "Lord of the Ice" belonged to history. That his status as head of operations served as little more than a courtesy title was gradually brought home to Siple as he observed a poorly disguised relegation of his old commander to the back seat.

> There was in some quarters of the Navy Department, I learned, resentment of Admiral Byrd partly because his various successes and the publicity he had amassed, partly because his spectacular promotions (he had risen from commander to rear admiral) had come without his having been in the active list . . . Then, too, Byrd had powerful friends in Congress, as well as being a personal friend of President Roosevelt. Yet in this quiet war, I now found that I as a close associate of the Admiral was also a target.[15]

After four Little America expeditions, Paul Siple perhaps knew better than most the complex and enigmatic personality behind the "ice lord" who had raised the American flag on modern Antarctica.

The erratic nature of a man troubled by doubts over his exercise of authority, and sometimes close to paranoid in his fears of disloyalty and insurrection, had been brought home to this awestruck Boy Scout at Little America I. In secret Byrd called a chosen few aside and announced that he had decided to admit them to his "Loyal Legion." To this purpose he administered an oath that went, "I solemnly swear on my word of honor and by all that I reverence . . . that whenever you call for my assistance in the name of the Loyal Legion . . . I will take such action as you request . . . that I will not divulge the request to anyone; that in case of disloyalty displayed in a crowd when you are present, I will act in response to a predetermined signal and predetermined course of action . . . To all of this I swear so help me God."[16]

Eight years after Highjump folded its tents, American forces once again stormed Antarctica under the banner of Operation Deep Freeze—a massive mission of peace. By icebreaker, aircraft, and conventional vessel came men by the hundreds to pursue the scientific programs of the 1957–1958 International Geophysical Year. Supply ships led by *USS*

Prelude to Highjump: The U.S. Coast Guard icebreaker Northwind *calves out a basin to allow supply ships entry into the Bay of Whales.*

Glacier, at 8,000 tons and 21,000 horsepower, the Navy's most powerful 'breaker, fat-bellied C-124 Globemasters, P2V Neptunes, and R4D Dakotas in the summer of 1955 began heading southwards from New Zealand towards "the ice."

Scientific and support bases were to be erected on the Ross Sea coast at McMurdo and (jointly with New Zealand) at Cape Hallett, at Byrd on the Rockefeller Plateau on 80°S and, in a major air drop exercise, at the South Pole itself. Continuing the tradition set by Admiral Byrd, yet another Little America would arise at the eastern end of the Ross Ice Shelf. However, a report that Siple had written after his previous Antarctic stay, aimed at explaining the gargantuan forces at work within the Bay of Whales, proved dramatically accurate.[17]

> Within four or five years the navigable portion of the bay will be completely compressed and by that time a great cataclysm will be close at hand which will cause much larger portions to be broken out under these irresistible forces meeting. The west side may continue to break for years to come, for it appears to be the more easily dislodged, but the east side will have to give also. The cycle of the Bay of Whales cannot be determined with certainty as yet but it is probably about fifty years. When the great cataclysm occurs, the sites of Little America I, II, and III as well as the more precarious tent camp of Operation Highjump will float out to sea.

The icebreaker *USS Atka* with helicopter support was charged with checking the suitability of the Bay of Whales—if the Bay still existed? They found a ten-mile (sixteen kilometer) strip had separated from the Barrier, and the Bay of Whales as previous explorers knew it had disappeared. *Atka* was able to navigate farther south than an American ship had reached before, though still eleven miles (seventeen km) short of *Fram*'s position in 1911. Inevitably, as Siple predicted, the east and west headlands had collided and ice cliffs 140 feet (45 m) now barred the way. Little America V would have to move elsewhere.

When planning for Operation Deep Freeze began in the early 1950s, the disregard for Byrd in certain senior Navy ranks grew even more apparent. Antarctic Task Force 43, under Rear Admiral George Dufek, completed seven of its twelve major flight assignments from McMurdo Sound without any reference to the commander. Siple recounted a particular episode when the Admiral, now in his late sixties, was relegated to a second-rate, uncleaned cabin in the lower decks of the *USS Glacier* for a voyage across the Ross Sea. Byrd behaved as if disrespect was of no consequence and rather a judgment on those who practiced it. When Siple moved to improve his accommodation, he replied—"My peers will never insult me; my inferiors cannot insult me."[18]

The call of Antarctica brought Paul Siple back again on his sixth and greatest polar adventure. His appointment made him first scientist-leader in 1957 of Amundsen-Scott Station at the South Pole. That men and women might one day occupy a permanent settlement at 90°S would be a venture beyond the wildest dreams of those heroic conquerors of the Pole whose names the Station bore. "We are living at the bottom of the world," said Siple in a *Time* magazine interview that featured his photograph on the front cover.[19] Preparing for the Pole operation, he recalled living in a tent at McMurdo Sound close to Scott's famous Discovery hut of 1902, and reflecting that this was the seventh Christmas Day he had spent in Antarctica.

"The ice" earned Paul Siple international fame at the same time as Admiral Byrd, now in his seventieth year and denied the helm of Antarctic planning, faded from the scene. He suffered heart failure and died at his home in Virginia on 12 March 1957. Siple received the news in a message radioed to the South Pole.

A conversation with Siple soon revealed his continuing loyalty to the leader who had first opened for him the doorway to Antarctica. Though probably not always liking what he saw, his inner feelings were guarded and rarely disclosed. Anyway, his book bearing the simple title

90° *South* tells his life story—as much as he wants you to know it. I have a valued, autographed copy that he handed to me in his Canberra office one day before he said farewell.

Paul Siple returned to the United States where, aged only fifty-nine, he died of a heart attack on 25 November 1968, survived by his wife and daughters.

One sails the Ross Sea in the hope of a rendezvous with a slice of Antarctic history riding on a passing iceberg. Framheim and the Little Americas have floated to a watery grave with the inevitable calving of the Barrier ice on which they stood. But you never know. In 1963 a lookout on *USS Edisto* sighted a 'berg at 77°32'S with the remains of a camp visible beneath the top layers of snow. When Little America IV was hastily evacuated in 1948, nine valuable aircraft of Operation Highjump were left behind. A few years later they were not to be seen at the Bay of Whales.

Maybe it's too late now, but what fascination to meet Dr. Thomas Coulter's mammoth Snow Cruiser passing by on the longest Antarctic trip it would ever make.

A three-point landing minus pilot: The Norseman scout aircraft is lowered to the bay ice from flagship USS Mount Olympus. *Operation Highjump was designed as a polar rehearsal for U.S. military forces, in the event of war breaking out across the Arctic zone.*

The Snow Cruiser was one of the more spectacular blunders of the American polar program. Designed as nothing less than a self-contained and mobile base, it was intended to drive all the way to the South Pole, bridging crevasses, tobogganing on its belly down snow slopes and sending the small Beechcraft aircraft fastened to its roof away on scouting missions.[20] The Snow Cruiser was so large that it had to be built, at a cost of $150,000 (in 1930s terms), at the Pullman railroad company's workshops. A laboratory, galley, office, sleeping quarters, radio centre, darkroom and workshop were contained within the 35-ton metal body, which measured 55 feet long, 20 feet wide and 15 feet high.

But when delivered to Dr. Siple's West Base in 1940, the Snow Cruiser's problem was that it couldn't pull itself along. The ten-foot-diameter wheels, which weighed three tons each, dug a hole in the snow from which the diesel electric motors lacked sufficient power to move them on. The Cruiser barely climbed the slope from the Bay of Whales to Little America III, where it expired and never moved again.

No lack of power troubles *Kapitan Khlebnikov*. At 12,288 gross registered tonnage, we are voyaging on the most powerful of Russia's conventional icebreaker fleet (only the nuclear ships are stronger). In an engine room so clean "that you could eat your dinner off the floor" (to quote Terry, one of our fellow travelers), the six big Wartsila diesel-electrics thunder out 24,000 horsepower—all of which may be needed as we close with the Barrier in a search for Farthest South. Three huge propellers 4.3 meters (14 feet) in diameter each with four hardened steel blades churn an ice-littered wake. Cruising speed is sixteen knots and full speed can take us to nineteen knots. Thickness of the double hull is forty-five millimeters at the ice-breaking section and an "ice knife" is fitted beneath the distinctive spoon-shaped bow. To augment the icebreaking capacity, pumps move seventy-four tons of water a minute between ballast and heeling tanks, and a row of tiny ports (an "air curtain") extending forward

to midships about six feet (two meters) above the keel, pump compressed air outwards from the hull against the pressure of the ice.

As we smash our way towards the Bay of Whales, one wonders how little ships like *Wyatt Earp* really managed it.

Beside Scott's 1902 Discovery hut on Winter Quarters Bay, Paul Siple sheltered in a tent at the start of Deep Freeze in 1955. Modern McMurdo base now looks across the historic site with USCG Polar Sea *in the Bay.*

On the ice edge of the Bay of Whales travellers, lowered from the ship in a metal cage, gather to toast Kapitan Khlebnikov's *arrival at Farthest South.*

CHAPTER SEVEN

END OF THE WORLD

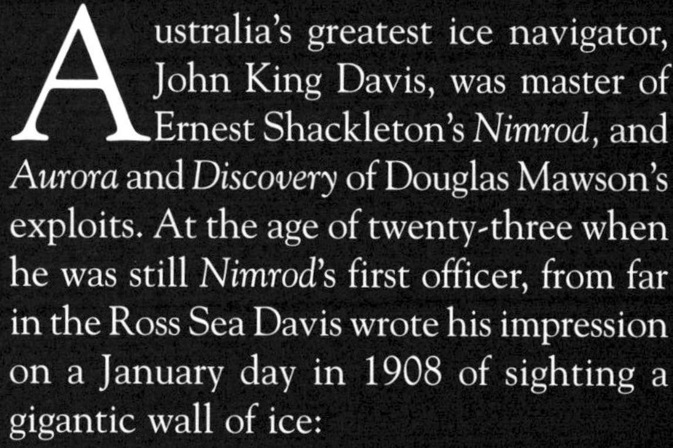

Australia's greatest ice navigator, John King Davis, was master of Ernest Shackleton's *Nimrod*, and *Aurora* and *Discovery* of Douglas Mawson's exploits. At the age of twenty-three when he was still *Nimrod*'s first officer, from far in the Ross Sea Davis wrote his impression on a January day in 1908 of sighting a gigantic wall of ice:

> On the 22nd all eyes were strained to the southward where, sure enough, we saw the white reflection in the sky, the ice blink, that we knew must

betoken our goal. On the morning of the 23rd, a hard white line, as if drawn with an enormous ruler, began to materialize out of the blink, becoming sharper and more clearly defined as we drew near, until at last it stood fully revealed as the Great Ice Barrier itself.

We had arrived. And from the blue and white face of these tremendous ice-cliffs there flowed toward us a current of air having in it a quality of coldness that was entirely new to me. It was the wind of the Antarctic, blowing to us straight from the heart of the last continent.

The cliffs of Dover turned to ice seemed to confront us at the other end of the world. We steamed in close under the vertical face of the Barrier. Now the chill wind passed overhead, the sea was calm and the sun was reflected in every colour of the spectrum from the countless facets of the ice and snow. As we held our course to the eastward along it we were astonished at the extraordinary uniformity of the level upper edge of the Barrier that towered, like the coping of a modern sky-scraper, high above us. Below, we looked deep into gigantic caves in the ice that could have contained the ship herself, huge, echoing cathedrals hewn out of sapphire and emerald and floored with the purple carpet of the eroding sea.[1]

Aboard *Kapitan Khlebnikov* we are sailing the same course that *Nimrod* followed when Shackleton searched for a base to locate his British Antarctic Expedition. However, our experience is different, for the wall of ice we track is, in truth, not the Barrier. It is B15, possibly the largest iceberg the world has ever known.

In March 2000 this behemoth of the south separated from the face of the Ross Ice Shelf. Whether it finally broke free with a thunderous roar of collapsing cliffs and avalanching ice or with barely a whimper we will never learn. Satellite images only report on B15's size and whereabouts; they do not contain a sound track.

At its calving, scientists from the University of Wisconsin calculated B15 covered an area of about 11,000 square kilometers

(4,500 square miles) and held enough fresh water (528 trillion gallons) to satisfy the entire needs of the United States for several years.[2] Its length measured 190 miles (300 km), equivalent to the distance from Boston, Massachusetts, to New York City; and below water level it reached for 700 feet (220 m) and more into the deep.

"Iceberg" seems a term inadequate to describe this "godzilla of the poles," as one scientist said in describing its capacity to wreak untold damage. Shortly after calving, B15 collided with the eastern Ross Ice Shelf, splitting off other massive 'bergs numbered B17 and B18 and several of a "smaller" size. Emperor and Adelie populations are put at risk when these icebergs go aground, possibly denying them access to the normal penguin feeding waters.

The blue-tinged tabular 'berg that has been to our starboard since early this morning is actually "B15A," which resulted after two months of tide and movement cracking apart the original monster, which now drifts in several separate sections, each one the size of a city or a state. Researchers from University of Wisconsin flew in a helicopter from the Coast Guard vessel *Polar Sea* to place weather recorders and Global

Amundsen's Fram *sailing along the Great Ice Barrier. The famous Norwegian vessel reportedly touched 78°40'S in making a departure circuit of the Bay of Whales.*

Positioning instruments on B15A to check its movements. Later we will take to our helicopters and land on top of B15D to frolic about on the thick snow, building an igloo and raising the flags of several nations. This is a sport few are given to enjoy. (The "B" nomenclature belongs to the Ross Sea sector, and "15" relates to the order of discovery).

The Ross Ice Shelf from which B15 calved advances northwards at the rate of about five feet daily, say a meter and a half. Prior to the B15 event, the face of the Shelf—which historically we call the Great Ice Barrier—was estimated to be about thirty miles (forty-eight km) north of previously calculated latitudes. The Shelf itself covers some 160,000 square miles, equal in area to the size of France or California. Limitless ice feeds towards the calving board—ice generated from Marie Byrd Land and the lofty polar plateau, from the very Pole itself; through the yawning glacier valleys amid the Trans-Antarctic mountains,

Right: From the bridge, New Zealand explorer Dr. Harry Keys charts the icebreaker's far southerly course as it cruises the Great Ice Barrier. Below: Subsidiary 'bergs broke from gigantic B15 on its journey into the Ross Sea. Travellers from Kapitan Khlebnikov *landed by helicopter on top of an immense section known as B15D.*

the Beardmore, the Shackleton and the Liv, ice is ever marching towards the sea.

Is there a sinister import in the calving of ice giants such as B15? Is global warming releasing the white outriders of a continental torrent that one day will make Sydney, London, New York, and Tokyo the new Atlantis of the twenty-first century? Other recent spectacular breakouts have been noted, from the Larsen Ice Shelf, the Ronne, and the Pine Island. And in May 2002, two years after B15, another huge 'berg, classified C19, measuring 125 miles (200 km) long and 20 miles (32 km) wide, separated from the Barrier's western end. Will the Archimedes principle keep the world in balance? Will ocean warming provide the ultimate trigger? Mostly the experts seem to shrug their shoulders. A dramatic shrinkage of the extended ice shelf seemed inevitable. Icebergs are a fact of nature. Since the dawn of the ice age they have been calving and going to their graves, some drifting north, some aground. Satellite images will continue to yield their message. Keep watching this space.[3]

Left: An almost "Farthest South" reading on the GPS monitor located on Khlebnikov's *bridge: 78°37.46'S and 165°02.3'W.*

Seven hours it has taken us to sail the 106-mile (170 km) length of B15A. The kaleidoscope of sheer cliffs pock-marked with caverns and crevasses, walls once glassy smooth, next shot with crazed geometrical patterns, comes to an end. Our eastwards course turns to the southeast. Between B15A and the new face of the Barrier, a narrow channel exists, but we dared not risk a short cut for to be seized between the 'berg and the Barrier would consign our 13,000 tons of tough steel to the swift fate of an egg beneath a sledgehammer.

When the Americans returned on *USS Atka* in 1954 to seek a site for Little America V in the forthcoming Deep Freeze program, the steep cliffs of the Bay of Whales refused them entry. Instead they turned towards an alternative location of Kainan Bay, named after the Japanese expedition ship of 1912, which lay thirty-five miles (fifty-six km) to the east. The Bay of Whales, through the forty or more years of accepting man's intrusion, had also let man off very lightly, considering the ever-present dangers. But at Kainan Bay disaster did not delay.

Father Daniel Linehan, the eminent Jesuit seismologist—the man who first depthed the ice at the South Pole—was flying to join a shore survey party in *Atka*'s helicopter piloted by Lieutenant John Moore when the world around them suddenly dissolved into a treacherous milky whiteout that gave no clue to height, angle or depth. "By luckily locating other men on the ground, we were able to land safely on shore," remembered Father Linehan.

> When John left me there and started back to the ship, I told him to get up very high on take-off so he could see the ocean and use the horizon as a guide. He took off all right, but as I was putting on my skis I heard him coming back. He'd lost sight of the ocean and thought he was flying level. But I could tell he was flying straight for the ice. There was a tremendous crash, and I ran to the wreckage and

cut him out. The stick had gone right into his stomach. He died twenty minutes later.

If all Antarctic travel contains an element of the unexpected, in aviation—as evidenced in the loss of helicopter pilot John Moore—the odds of disaster are perceptibly higher. Which brings me to recall making the first flight from Australia to the South Pole, barely beyond the winter darkness of 1964; the opening chapter of my Antarctic aviation history, *Moments of Terror,* reminds of a journey that was in danger of going horribly wrong.

> I remember holding a piece of cardboard over my head and trying to bury myself among the survival kits and sleeping bags. The crew knew the drill better; they had pulled on hard hats and belted themselves to the rear seats. It was the end of the first direct flight from Australia to the South Pole, and we were coming in for a crash landing.
>
> We weren't landing at the Pole itself. It was far too cold for us to land our seventy-ton Hercules at the bottom of the world; we'd never get off again. Minus 67°C (–90°F) when we opened the hatch for dropping the mail and newspapers to the twenty-two men living in blizzardly isolation at Amundsen-Scott Station. Nor were we at McMurdo Sound, site of the main United States base in Antarctica, though it was supposed to be our ultimate destination. Williams Field (named after the first American to die at McMurdo) had been ruled out, reporting an eighty-five-knot crosswind on the skiway and failing visibility. Instead we were headed 900 miles (1,440 km) across the wilds of the Rockefeller Plateau, making for Byrd Station, the Americans' "city of science" hidden beneath the ice cap.
>
> Normally, no one in their right mind would want to fly to Byrd with winter barely over. Can we make it? Lucky, if we do.

Not because there was anything wrong with our original fuel and weight computations; the U.S. Navy crew led by Commander Fred Gallup, who heads the VX-6 Antarctic aviation detachment, had been doing their sums for the past four months, practising for this day. But planning presupposed a regular 26,000 foot (8,000 m) cruising altitude (give or take a few thous), and no crazy detours. Fuel reserves now are running low, the mountain crags seem barely beneath our wings—we're flying unpressurised.

Did I tell you how we lost cabin pressure? Why we began to go wobbly at the knees, talk funny, and me to write all squiggly in my reporter's notebook? The culprit is that mail drop over the South Pole a few minutes before. The side hatch was open only for about thirty seconds, but long enough for rime ice to form around the frame, instant condensation when the heated air inside the cabin met the cold blast outside. The hatch seemed to close, but it didn't, not for the last few centimeters. Leaking pressure, the Hercules climbed back through 18,000 feet (5,500 m), which is when things turned nasty.

Emergency procedure on the flight deck. Fred Gallup (thank heavens he was breathing oxygen) smartly brought us down to a level where we didn't want to vomit; an oxygen cylinder was broken out and, like members of an oriental opium gang, we took turns to suck from the mask. Just about this time, one of the crew started crawling on all fours, not gone giddy but purposefully attaching a cargo chain to the drop hatch and passenger door.

We were told to belt up and stay put until he completed the exercise. With the pressure change, there was just a chance that the openings might suddenly pop and all of us could be sucked out . . . as sometimes happens to James Bond's enemies. The snowy surface we've been flying above for the past seven

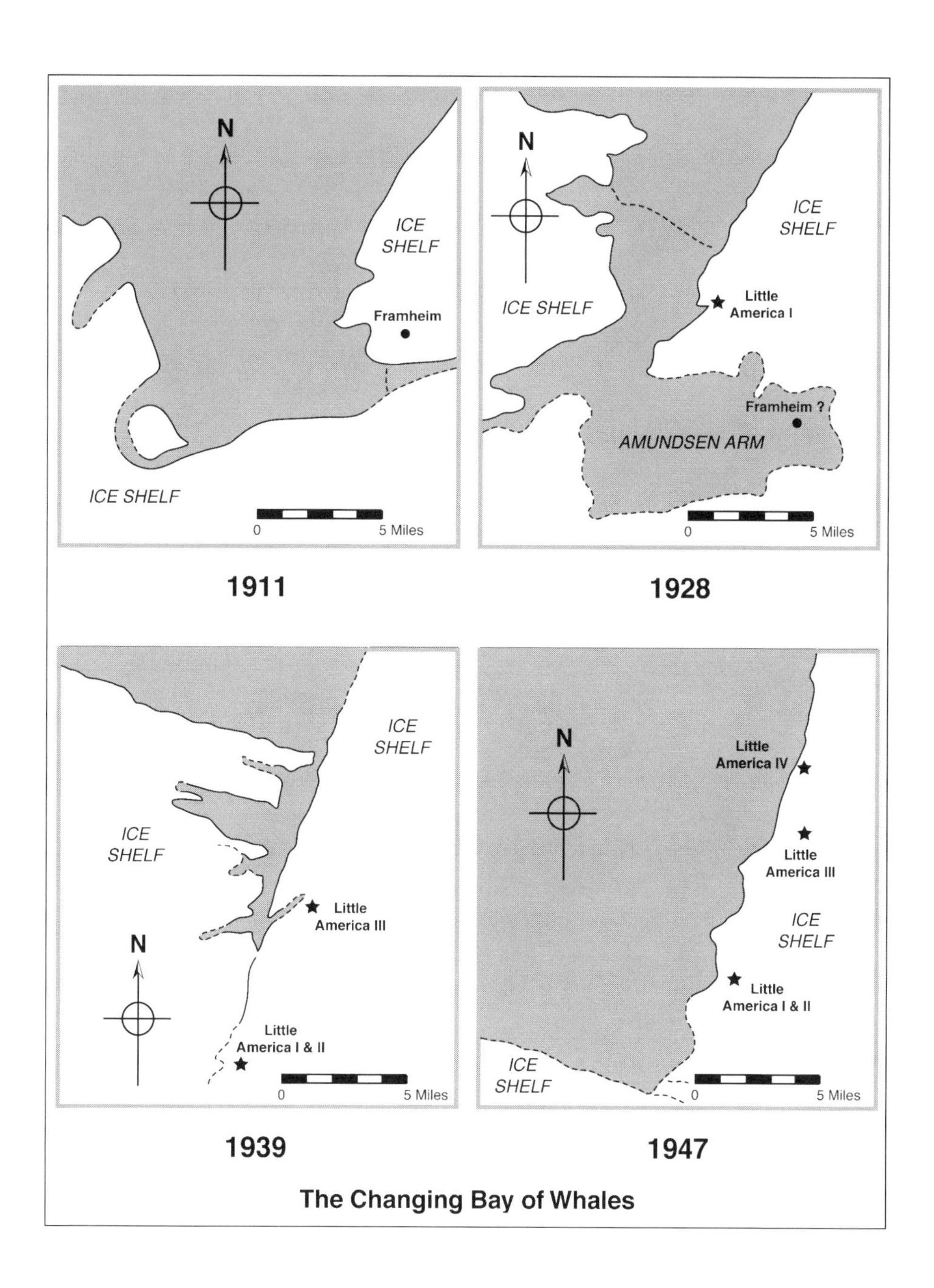

The Changing Bay of Whales

hours might look white and soft from aloft, but it's not. It is creased with diabolical sastrugi—icy ridges hardened by the wind—and scarred with blue and sinister crevasses. No place for a free fall. No place to land. No place to live.

We level out and are approaching the vast chain of mountains that guards the rim of the polar plateau. If we're at 12,000 feet (3,700 m), some of these unfriendly fangs must be just as high—and maybe more. We have a close-up view of every rock-faced peak, of every sheer ice fall and of the vast serrated glaciers that wind through the ranges, one day to reach the edge of the Great Ice Barrier where the mighty tabular 'bergs are born. Fine view, but no place to land. Half an hour away from the Pole, our course is changed for Byrd Station. We're flying farther and lower, and using up the fuel reserves. Worse is to come.

The "great cataclysm" that Paul Siple forecast has happened. The Bay of Whales lies in ruins. Ice in turmoil. Ice torn apart. Shattered walls. Cracks and crevasses revealed as if at the hands of a maniac sculptor. Not the Bay that Byrd and Amundsen sailed. All of it laid waste with an utter indifference to man and his puny works.

Captain Petr and his officers scan the ice front for a landing place. Excited travellers crowd the bridge, prompting leader Werner to call for shush. Some serious navigation is afoot. Harry, our New Zealand glaciologist, leans across the chart, tracing by radar, binocular, and eyeball a coastline (if such a term can be used) that no one before us has ever seen. He estimates the cliffs average 150 feet (46.5 m) high. Names are bestowed on new features—Julian Point, Quark Inlet, Grigory Head, Petr Gulf, Khlebnikov Bight; one mighty slab of ice is named after the author—David 'berg. Beat that.

Travellers cluster around the Global Positioning display. 78°25 . . . 78°30 read the little dancing numbers. Captain Petr calls an order in

Russian to his helmsman and *Khlebnikov* begins a ninety-degree swing, our bow aimed at the edge of the sea ice. Through uncertain greyish light the Barrier itself beckons in the distance. Diesel-electric controls are pushed full ahead. Down in the engine room thunder is let loose.

Khlebnikov lunges forward. Ice splinters. Ice cracks. Ice meters thick falls apart in huge blue and brown-stained chunks, to disappear in a swirl beside our stout hull. Not good enough, Captain Petr decides. Controls thrown into reverse. We can take comfort that the three propellers and rudder are protected by a clever "ice horn" device. Little chance in going astern of ice damage to propulsion that frequently dogged ships of the mechanical age, old and new.

Preparations are made for landing on the ice. Dropping the gangway is too risky. Transport will have to be in a metal cage, eight travellers at a time, lowered by the ship's forward crane. Passengers will have second priority. Trestles, glasses, and champagne come first. But the toast must wait—we haven't quite got there yet.

The end of the world is nigh! In we come again, *Khlebnikov*'s huge spoon bow rising and dropping, thrusting ice apart. 78°36 . . . 78°37 flick across the GPS window. We have voyaged 2,500 nautical miles across stormy seas and ice calm for this moment. Another full ahead

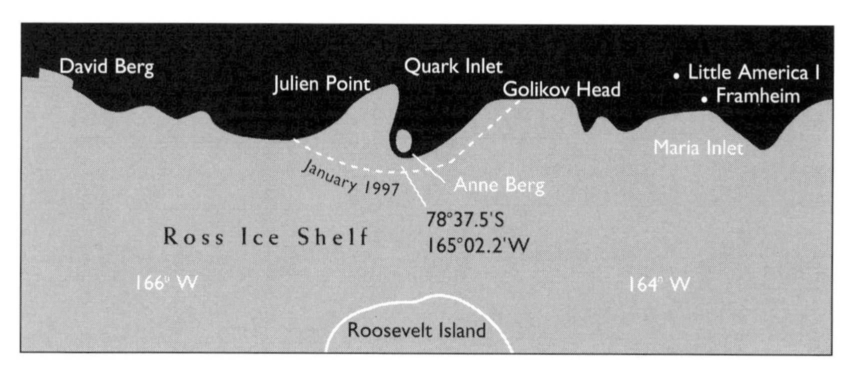

A copy of Dr. Harry Keys' chart of the Bay of Whales, January 2001

assault, until the ship is shuddering in the face of impassable ice by the trillion tons. Enough, says Captain Petr. At two o'clock in the afternoon we halt, our proud Russian icebreaker firmly wedged into the icy remains of the Bay of Whales. Looks of relief on Captain Petr and his crew. Werner makes an important announcement. Everyone claps. Congratulations to the Russians. We are at Farthest South a passenger ship has ever docked in Antarctica. Farthest south of any ship since the time of Amundsen and his *Fram* ninety years before—78°37.5'S.

The South Pole is a mere 682 nautical miles away.

THE BARRIER SILENCE

The Silence was deep with the breath like sleep
As our sledge runners slid on the snow,
And the fate-full fall of our fur-clad feet
Struck mute like a silent blow
On a questioning "hush," as the settling crust
Shrank shivering over the floe:
And the sledge in its track sent a whisper back
Which was lost in a white fog-bow.
And this was the thought that the Silence wrought
As it scorched and froze us through,
Though secrets hidden are all forbidden
Till God means man to know,
We might be the men God meant should know
The heart of the Barrier snow,
In the heat of the sun, and the glow
And the glare from the glistening floe,
As it scorched and froze us through and through
With the bite of the drifting snow.

—*Dr. E.A. (Bill) Wilson*
Written at McMurdo Sound, 1911.[6]

January 11, 2001
News Release

The Russian icebreaker *Kapitan Khlebnikov* has today reached a new southernmost record for shipping in Antarctica at 78 degrees 37 minutes south. This was in a new part of the Bay of Whales, which has been totally reformed by the calving of the giant icebergs B15, B17, and B19.

Passengers were landed and walked over the fast ice along a flagged route inland and onto the Ross Ice Shelf.

The new icefront of the Shelf formed by the calving of B15 has been mapped by scientists and officers on board. Earlier, the ship sailed past the B15A iceberg, taking seven hours to traverse its huge length of 170 km.

Petr Golikov
Master
I/B *Kapitan Khlebnikov*

Werner Stambach
Expedition Leader
Quark Expeditions

Appendix: Kapitan Khlebnikov

With the turn of the 1990s, the availability of the former Soviet's fleet of icebreakers and ice-strengthened ships, coupled with Russia's need for Western currency made accessible to the tourist industry such powerful ships as *Kapitan Khlebnikov,* from Vladivostok.

Under the command of Captain Petr Golikov, a veteran of thirty-seven years in polar seas, and with a highly experienced Russian crew totaling sixty men and women, and leased to the Quark tourist organization of Connecticut, USA, *Khlebnikov* began a series of cruises across the southern seas, carrying some ninety passengers and a management group of expert polar commentators and guides.

Helicopter- and Zodiac-equipped, *Khlebnikov* was able to penetrate the ice-bound shores of Antarctica in the historic Ross Sea region, which formerly had been open only to the U.S. Coast Guard icebreakers.

Kapitan Khlebnikov, built in 1981 in Finland and of 12,000 gross tonnage, is powered by six diesel-electric motors and totalling 24,000 horsepower, enabling the ship to maintain cruise speed of sixteen knots in open water. *Khlebnikov* is classed amongst the world's powerful icebreakers, class KMxLL3A2.

Specifications

Length:	132.40 meters
Breadth:	26.75 meters
Draft:	8.50 meters
Gross Tonnage	12,228 tons
Machinery:	Six Wartsila Diesel electric 24,000 total h.p.
Displacement	10,471 tons
Cruising speed:	15 knots
Crew:	60
Passengers (max.):	112
Registry:	Russia
Classification:	Full icebreaker class

Captain Yuri Khlebnikov (b. Warsaw 1900) was a distinguished Russian ice mariner who served in Arctic seas for thirty-two years until his retirement in 1969.

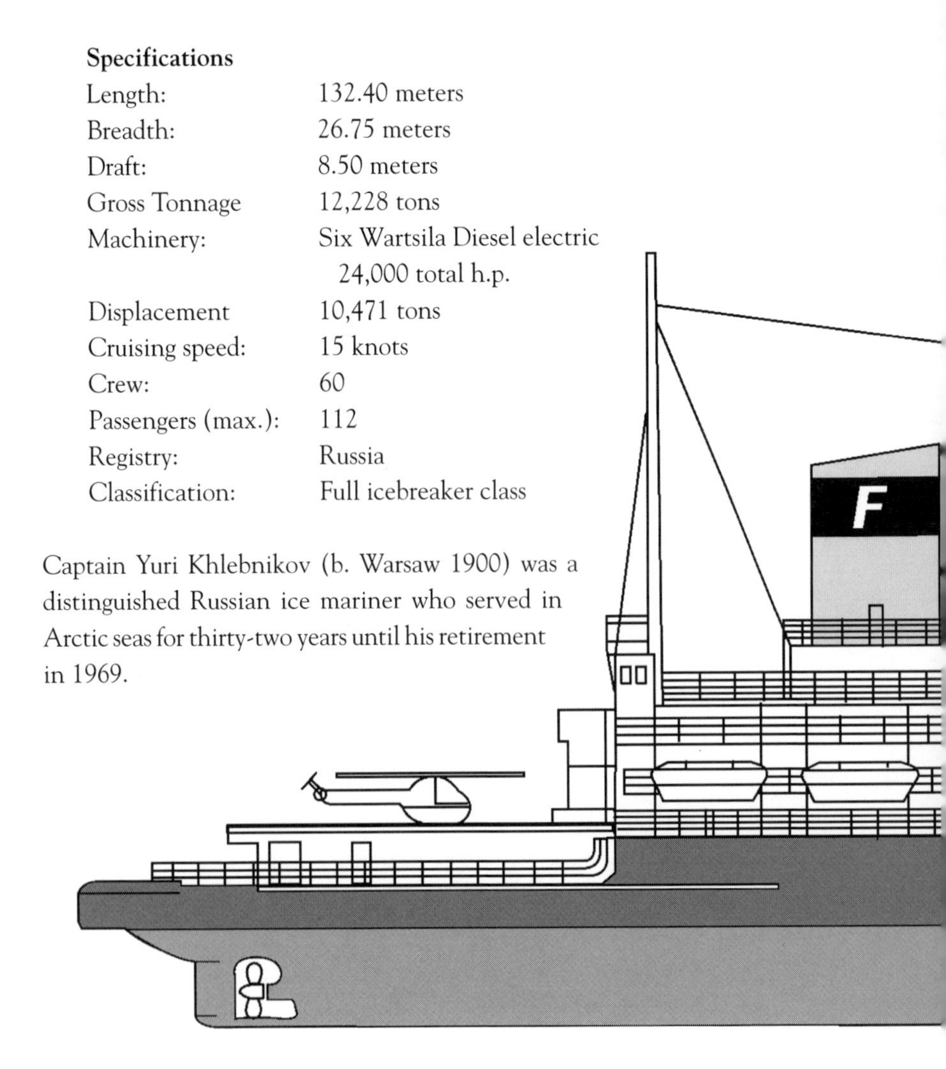

Registry

Russia. 2 helicopters on board. Fleet of Mark V heavy duty zodiacs.

CAPTAIN: Petr Golikov

SECOND CAPTAIN: Gennadiy Antokhin

FIRST CHIEF MATE: Andrey Gostnikov

SECOND CHIEF MATE: Sergey Sorochanov

CHIEF ENGINEER: Aleksandr Stepura

CHIEF ELECTRICAL ENGINEER: Valeriy Bogach

SHIP'S PHYSICIAN: Dr. Oleg Vovk

CHIEF STEWARD: Igor Belozerov

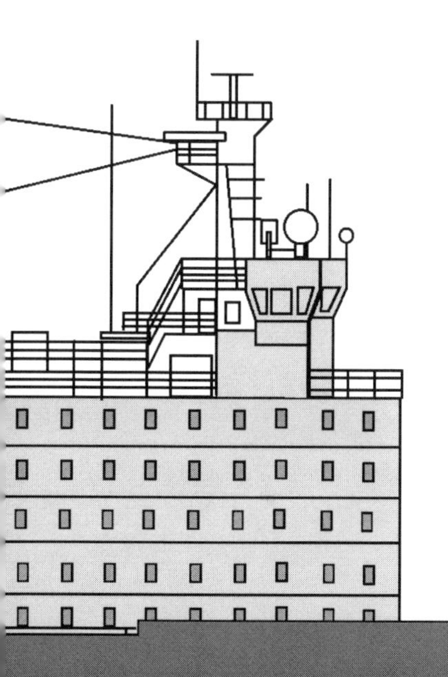

CHIEF RADIO OFFICER: Vasiliy Tatarnikov

HYDROLOGIST: Anatoliy Moskalyov

Vladivostock Air Helicopters

CHIEF PILOT: Sergey Sirak

PILOT: Anatoliy Komysh

PILOT: Nikolay Tkachenko

PILOT: Aleksander Belyy

ENGINEER: Evgeniy Sukhikh

AVIATECHNIK: Vitaliy Rogov

AVIATECHNIK: Oleg Egorov

Expedition Staff

NZ GOVERNMENT REPRESENTATIVE: Rob McCallum

LEADER: Werner Stambach

ASSISTANT LEADER: Susan Adie

ASSISTANT STAFF: Belinda Sawyer

NATURALIST: Brad Stahl

NATURALIST: Luke Saffigna

HISTORIAN: Bob Headland

GLACIOLOGIST, GEOLOGIST, VULCANOLOGIST: Dr. Harry Keys

ORNITHOLOGY: Nigel Milnius

MARINE BIOLOGY: Keith Springer

PHYSICIAN: Dr. Ian Rogers

Farthest South

Ship	*Date*	*Position given at or near Bay of Whales*
Erebus / Ross	2/3 Feb 1842	78°04S
Southern Cross / Borchgrevink	17 Feb 1900	78°34.37S, 164°32.45W
Discovery / Scott	26 Jan 1902	78°36S[1]
Nimrod / Shackleton	22/24 Jan 1908	78°20S, 162°14W
Fram / Amundsen	15 Jan 1911 15 Feb 1911	Docked – 78°38S Sailing out – 78°41S
City of New York / Byrd	2 Jan 1929	78°34S[2], 163°56W
K. Khlebnikov[3]	See under 11 Jan 2001	 78°37.5S, 165°02.03 W

1. Between the *Erebus* and *Discovery* voyages, comparison of their charts seems to indicate the Barrier had receded up to thirty to forty miles in places from previous position.
2. Little America I is also mentioned at "four miles north of Framheim." All future Little Americas were located farther north again.
3. Latitudes unknown of other vessels reaching Bay of Whales: 1934–1936 *Wyatt Earp* and *RRS Discovery II*; 1947 Operation Highjump; 1955 Operation Deep Freeze. However, upon consultation with Dr. Harry Keys (NZ), it seems the position of these ships did not exceed 78 degrees.

End Notes

ABBREVIATIONS

Advertiser	*Adelaide Advertiser*
Geogr. Rev.	*Geographical Review*
Nat. Geo.	*National Geographic*
NYT	*New York Times*
SMH	*Sydney Morning Herald*
Times	*The* (London) *Times*

The author's *Moments of Terror: The Story of Antarctic Aviation* contains the full narative material of extracts appearing in chapters 3, 4, and 5.

FRONT MATTER

1. Born in 1883 in Bowral, N.S.W., Frank Debenham was the son of a Church of England minister. After graduating as a geologist at Sydney University,

he joined Scott's 1910–1913 British Antarctic Expedition. After serving in World War I, where he was badly wounded, he went on to become first Professor of Geography at Cambridge University and a founder of the Scott Polar Research Institute. He died in 1965.

CHAPTER 1

Quotations referenced in this chapter are taken from the following: Borchgrevink, C., *First on the Antarctic Continent;* Debenham, F., *Antarctica: The Story of a Continent;* Ross, M. J., *Ross in the Antarctic;* Scott, R. F., *The Voyage of the Discovery;* Shackleton, E., *The Heart of the Antarctic;* (see Bibliography for complete entries). References to the voyage of *Kapitan Khlebnikov* are taken from the author's notes.

1. Ross, 93–97. The epic voyages of Sir James Clark Ross, RN, in *Erebus* and *Terror,* 1839–1843, resulted in many discoveries, the most memorable of which are Victoria Land, Mount Erebus, the Ross Sea, and the "Great Icy Barrier."

2. Borchgrevink 277–279. Born in Oslo in 1864, Carstens Borchgrevink in 1899 led the first Australian expedition to "winter over" on the Antarctic continent at Cape Adare.

3. Scott 146–148. At the age of thirty-three, Robert Falcon Scott led the National Antarctic Expedition of 1901–1904 to McMurdo Sound. His sledging party with Shackleton and Wilson reached a record 82°S.

4. Shackleton 45–52. In 1907 Shackleton returned to Antarctica with his own expedition, which was notable for the first ascent of Mount Erebus, the location of the South Magnetic Pole, and an attempt to reach the South Pole.

5. Debenham 112, 116, 209. Frank Debenham, born in 1883 at Bowral, New South Wales, was the first native Australian scientist-explorer to serve in Antarctica.

CHAPTER 2

Quotations where referenced are taken from Amundsen, R., *The South Pole;* Huntford, R., *The Last Place on Earth;* Martin, S., *A History of Antarctica;* Scott, R. F., *Scott's Last Voyage* (see Bibliography).

1. Hobart has many associations with Antarctic expeditions—historic and present-day. The hotel room where Amundsen lodged is now featured in the city's polar "tourist trail."

2. Born in 1872, Amundsen began studying medicine at the University of Christiania but upon his mother's death, decided to go to sea. In 1897 he served as mate with de Gerlache's *Belgica* scientific expedition which spent a full winter on the Antarctic Peninsula. In 1903, Amundsen became a nationally recognized figure when he completed the first successful navigation of the Northwest Passage in the forty-seven-ton sloop *Gjoa*. He next organized to drift across the North Pole in Nansen's ship *Fram* until learning that Peary had reached the Pole changed his plans.

3. Olav Bjaaland was the only expedition member who could sometimes joke at Amundsen's leadership. He was a champion Norwegian skier from Telemark who applied himself to perfecting the expedition's skiing techniques. He frequently skied ahead of the Pole party to urge the dog teams forward.

4. Nansen, who had provided Amundsen with *Fram* for the supposed North Pole attempt, expressed dismay that his fellow Norwegian had not confided in him his decision to try for the South Pole.

5. For six days of each week the Norwegians worked to a strict daily timetable throughout the winter of rising at 7:30 A.M., work at 9 A.M., lunch at noon, resuming work from 2 P.M. until 5:15. They ate well on a menu that ranged from buckwheat cakes to butter and cheese, seal meat and tinned meat, fruits and puddings. Brandy was permitted on Saturday evening after a spell in the sauna and aquavit on Sunday.

6. Fritdjof Nansen, a national hero, personified Norwegian achievements and independence. Aged twenty-seven, he led the first crossing of the Greenland Ice Cap in 1888–1889. His ship, *Fram*—purpose-built to withstand ice pressures—he next took on an Arctic drift and narrowly failed in an attempt to reach the North Pole. However, Nansen's emphasis on the use of skis and dog-team sledging showed the way for future polar exploration. A subsequent career in science, politics, World War I relief measures, and the League of Nations won him the Nobel Peace Prize in 1922. He died in 1931.

7. After the end of World War I, the Arctic drew Amundsen back on several expeditions—including a failed attempt with the American millionaire Lincoln

Ellsworth in 1925 to fly over the North Pole. The following year in the Italian airship of Umberto Nobile he finally succeeded in flying across the polar ice cap. When Nobile's dirigible *Italia* crashed in the Arctic in June 1928, Amundsen hurriedly flew out to find him and was never seen again.

8. Entry in Scott's diary for January 17, 1912—the day before attaining the South Pole. Also "there is this curious damp feeling in the air that chills one to the bone."

9. Priestly, who had served as assistant geologist to Professor David in Shackleton's 1907 expedition, felt irritated at the dismissals of his "boss's" achievements that were sometimes expressed within the Scott camp. His diary entry expresses his pleasure at seeing one important aspect of Shackleton's exploration vindicated. Sir Raymond Priestly later became vice chancellor of Melbourne University.

10. Wilfred Bruce joined Scott's expedition from the P&O merchant marine.

11. Shirase, a thirty-year-old Army lieutenant, returned to a hero's welcome, which was in marked contrast to the opposition he met at home in organizing Japan's first Antarctic expedition. However, the support of the prominent Count Okmuma enabled his plan to proceed. While awaiting their second attempt to reach the Bay of Whales, the expedition camped at Vaucluse on Sydney Harbour. The staunchly "White Australia" Bulletin lampooned the Japanese in a cartoon of 16 May, but they were befriended by Professor Edgeworth David, the Antarctic veteran from Sydney University.

CHAPTER 3

Quotations where referenced are from Byrd, R.E., *Little America;* Montague, R., *Oceans, Poles and Airmen;* Rodgers, E., *Beyond the Barrier with Byrd* (see Bibliography). Extracts are from the author's *Moments of Terror* and newspaper reports.

1. Bernt (Bert) Balchen, born 1900 in Norway, and later becoming an American citizen, served in Norway with the mounted artillery and as a flight lieutenant in the naval air service. He first gained prominence in polar aviation as a member of the Amundsen-Ellsworth-Nobile Arctic flight to Kings Bay in 1926. He first flew with Byrd in the 1927 Atlantic crossing, which reached the

coast of France. Balchen's impressive list of achievements included Middleweight Boxing Champion of Norway as well as being a noted ski racer. He was also known for his skills as a seaman, designer, mathematician, navigator, artist, cook, and carpenter. (He built a wooden lightweight man-hauling sledge in case the South Pole flight failed.) During World War II he helped to establish the Greenland air base and flew many dangerous missions to supply the Norwegian underground and rescue crashed airmen. He was president of Norwegian Airlines (parent of Scandinavian Airlines). He died in 1973.

2. On 21 December 1929, during his team's 2,400-kilometer (1,500-mile) sledging journey, Laurence Gould, Byrd's second-in-command, raised the American flag on Supporting Party Mountain in Marie Byrd Land and within a rock cairn, deposited a note stating "in the name of Commander Richard E. Byrd (we) claim this land as a part of Marie Byrd Land, a dependency or possession of the United States of America."

3. Byrd, 335–36.

4. Ibid., 340.

5. Montague, 261–63. Balchen noted that Byrd gave him no positions on the flight to or from the Pole. He regarded Byrd's mariner's sextant, to which was fixed an undampened carpenter's level bubble, as practically useless. Balchen also noted the smell of cognac on Byrd's breath when he approached the cockpit. Byrd had with him several pints of the liquid as treatment for a "heart complaint."

6. Byrd, 32–33. Byrd described actions taken at the Ford factory to increase engine horsepower and reduce the weight of *Floyd Bennett*. He had considered landing at the Pole, but became convinced that the aircraft's ski landing gear might collapse under the impact of rough wind-hardened snow on the polar plateau.

7. On 24 February 1934, to a note from the British Foreign Office querying the claims made by the Byrd and Ellsworth expeditions, the U.S. Assistant Secretary of State replied that his government did not propose to discuss the question and added, "However, I reserve all rights which the U.S. and its citizens may have with respect to this matter."

8. Rodgers cites Finn Ronne's 1979 biography *Antarctica My Destiny* (New York, Hastings House 1979), in which Ronne claimed that Byrd had confessed in

secrecy to Isaiah Bowman (President of the American Geographical Society) in 1930 that his famous 1926 flight had missed the North Pole by 150 miles (240 km). The Swedish meteorologist Professor Liljequist raised the same question in his 1959 paper. The account of the conversation with Floyd Bennett in Balchen's 1958 book *Come North With Me* was modified before final publication, reputedly under legal pressure from Byrd family interests. A feud that reputedly developed between Dean Smith, the expedition's second pilot, and Byrd after their return to America in 1930 was never resolved or properly explained.

9. Byrd station is no longer occupied.

CHAPTER 4

Quotations referenced are taken from Ellsworth, L., *Beyond Horizons* and Hill, L. C., *Antarctic Relief Expedition Report* (see Bibliography). Extracts are from the author's *Moments of Terror* and newspaper reports.

1. Ellsworth, 263–66. In farewelling *Wyatt Earp*, the Mayor of Dunedin said that Ellsworth's expedition comprised the finest group of polar adventurers that had ever visited the city.

2. Ellsworth's run of adverse luck continued. Worn gears caused through ice-breaking maneuvers made *Wyatt Earp* difficult to control at slow speed. As a result, the bow needed extensive repairs after the ship crashed against the wharf on reaching Dunedin.

3. Argus, 10 September 1934.

4. Wilkins spoke of a proposal to operate twelve weather stations around the Antarctic coast, observing the eleven-year sunspot cycle. He suggested an unnamed society was willing to invest one million pounds in the project if Australia contributed £250,000, however, the idea lapsed.

5. *NYT*, 19 March 1929. Montague, 259; Rodgers, 72–3.

6. *Geogr. Rev.* No 26, pp. 454–62, 1936. No 27, pp. 430–44, 1937. Ellsworth 294–9, stated that he was spending $5,000 a month to support his expedition.

7. Ellsworth, 360–61. Wilkins had made radio contact with *Discovery* and attempted a "race" to beat the British ship to the Bay of Whales for "the honor of the expedition." But when a sustained speed through 1,000 miles of thick pack threatened to damage the engine, "he desisted."

8. *Age*, 19, 26 December; *Argus*, 20 December; *SMH*, 26 November, 3, 6, 10, 12, 27 December 1935; *Times*, 31 December 1935; *Walkabout*, May 1936.

9. On 25 November, two days after Ellsworth's departure, Wilkins received a message from Kenneth Rawson, navigating officer of the second Byrd expedition, advising of food caches located at Mt. Grace McKinley and the Rockefeller Mountains, and at Little America where "there is ample food and coal in the tunnels."

10. Details of the Commonwealth Government's rescue mission, organized by Mawson and Davis, are held in the Antarctic file, Australian Archives, Canberra.

11. Ellsworth Antarctic Relief Expedition Report. *Discovery II*, a successor to Captain Scott's original vessel, was a steel-hulled motor ship of some 2,200 tons displacement.

12. The RAAF aircraft were A7-55, a DH60G Gipsy Moth, fitted with a 100-hp Gipsy engine and extra fuel tank to allow four hours' safe flying time. A5-37 was the Westland Wapiti, driven by a 550-hp Bristol Jupiter engine. Ski undercarriages were also taken.

13. A whiteout occurs when light is trapped between low overcast or cloud cover and the continuous surface of snow and ice beneath. As a result, horizon and shadow, and with them all perception of depth, altitude, and distance, are obliterated.

14. *Natone* sprang a leak while sailing between Lae and Brisbane and soon had more than a meter of water in the engine room. Sails were rigged and as she sought shelter in Rainbow Bay, grounded near Mudlow Rocks on the night of 24 January 1959. Waves smashed one lifeboat and when a thirty-degree list prevented launching of the other, the crew of eighteen drifted ashore on hatch boards, where a search party from Noosaville found them. *Natone* soon broke up and sank.

CHAPTER 5

Quotations where referenced are taken from Ellsworth, L., *Beyond Horizons*, and Douglas, G.E., in *Aircraft* magazine (see Bibliography). Extracts are from the author's *Moments of Terror* and newspaper reports.

1. Hollick-Kenyon was under age when he enlisted in the Canadian Expeditionary Force; he was twice wounded in France. Upon discharge, he joined the Royal Flying Corps. He did not get to fly in World War I, but in 1923 he joined Canadian Airways. By 1935 his 6,000 hours of flying experience included much time in the Arctic zone. "He was a fine fellow," said Ellsworth, "a grand pilot, and the quietest man I ever knew."

2. Ellsworth, 313–16.

3. Ellsworth, 324–25. The Pole lay 650 miles to the south, the coastline 450 miles to the north, and the Bay of Whales 670 miles ahead.

4. Ellsworth took a meteorologist on his first and second expeditions. Because Little America was now deserted, and thus no weather data could be exchanged, he excluded, with Hollick-Kenyon's agreement, a meteorologist from the third expedition. This vacancy also allowed space for the inclusion of the second pilot.

5. Ellsworth named the blizzardly Camp III after Hollick-Kenyon's home city of Winnipeg. Camp IV was called Tranquille, because of the prevailing windless conditions.

6. *Polar Star* carried a 300-watt HF radio and a petrol generator to supply power when landed on the ice; an emergency manual HF set was also included. Morse key was to be the main communication link with *Wyatt Earp* as distance increased.

7. Ellsworth, 334–35. By dead reckoning he believed they were only four miles short of Little America. They were, in fact, sixteen miles away, though their tramping would cover more than a hundred miles before they reached the old base.

8. Ellsworth divided his great circle route into fourteen sectors, each equal to one hour's flying. However, he was dismayed to find that instead of 240 kph (150 mph), *Polar Star* averaged only 163 kph (102 mph) across the continent. The excessive weight of the heavily laden aircraft (3,630 kg) would have been partly responsible for the lower performance.

9. To those who questioned his radio silence, Ellsworth replied: "Had our radio not failed, the world of the streets would have hailed the crossing of Antarctica as a most intricate and difficult undertaking in exploration carried through without a hitch. As to the charge that I suppressed the radio for the sake

of publicity, I would have had more publicity and the newspapers a more dramatic story than our mere 'disappearance' gave them, had I been able to send a daily account of our fortunes—how we fared during the long blizzard on Hollick-Kenyon Plateau, for example, or in our blind wanderings through the fog at the Bay of Whales."

10. Ellsworth *Antarctic Relief Expedition* report. Also diary of Flt Lt G.E. Douglas Aircraft, February 1986.

11. Other members of the RAAF detachment: Sergeants S. F. Spooner (engine fitter and emergency pilot), J. Easterbrook (metal rigger), J. W. Reddrop (wireless operator-mechanic), Corporal N. E. Cottee (metal rigger), ACI C. W. Gibbs (engine fitter).

12. *Age*, 17 February; *SMH* 18 February 1936.

13. Hollick-Kenyon returned to Canada, where he became general manager of Trans Canada Air Lines, Toronto, in 1938. He died in July 1975, aged seventy-eight; his name is entered in the Canadian Aviation Hall of Fame. At the rank of Group Captain, Eric Douglas retired from the RAAF after World War II and died on 4 August 1970, aged sixty-seven. His co-pilot became Sir Alister Murdoch, Chief of the Air Staff 1965–1970, he died on 23 October 1984. Douglas's son, Ian, led an ANARE party to Davis station in 1960.

CHAPTER 6

Quotations referenced are from Bertrand, K.J., *Americans in Antarctica 1775–1948;* Byrd, R.E., *Discovery: The Story of the Second Byrd Antarctic Expedition;* Dufek, G., Operation Deep Freeze; Siple, P., 90° *South:* Sullivan, W. *Quest for a Continent*, and the author's notes (see Bibliography).

1. Siple, 37. "All the finalists impressed me as being excellent Antarctic material . . . my advantage was that I was taller, heavier, and older than the others."

2. Siple, 59. I served as surveyor and navigator. From the top of Mount Grace McKinley we saw vast array of brick-red, black, gray, and brown mountain peaks . . . these had to be mapped because they had never before been seen by man. These mountains were only heads and shoulders poked up out of the ice flowing off the continental plateau rising 6,000 feet to the south and east.

3. Siple, 58. "I have known few men as brave as Admiral Byrd."

4. *NYT*, 21 February 1929, 12 January 1931. "I have claimed, in the name of the United States, 125,000 square miles of land we discovered," said Byrd. "(It) lies beyond the claims of Britain, being east of the 150th meridian." He added "Antarctica is big enough for all." However, Norway queried America's intentions, after Byrd's first expedition.

5. With a 267 kph (167 mph) cruising speed, the large Condor aircraft had a range of 1300 miles (2,080 km) with a 2,600 lb. payload. Alfred Sloan, the president of General Motors, loaned the Fokker, named *Blue Blade*, American Airways contributed the single-engine *Miss American Airways*.

6. Byrd, 113. "One saw taking form one of the most remarkable cities on the face of the earth—a city which would boast, among other possessions, of electric light and power, a complete broadcasting and field communications plant."

7. *SMH*, 13 January; 10, 17 July 1939; 6 March 1940.

8. Siple, 61–63. "The first I knew of the government's venture came in the spring of 1939, when Admiral Byrd called me. 'There is going to be a government expedition to the Antarctic,' he said, 'and I want you to take charge of the logistics of the expedition.' He added, 'We are planning three or four bases this time and you will be the leader of one.'"

9. The letter of instructions, signed by President Roosevelt on 25 November 1939, stated, "Members of the service may take appropriate steps such as dropping written claims from airplanes, depositing such writing in cairns et cetera, which might assist in supporting a sovereignty claim by the United States Government. Careful records shall be kept of the circumstances surrounding each act. No public announcement of such act shall, however, be made without specific authority in each case from the Secretary of State." In compliance with the presidential authorization, a U.S. seismic party on 12 December 1940 raised the American flag and left a statement of claim in a rock cairn at 78°06' South and 154°48' West, which would appear to be within the border of the Ross Dependency.

10. Three important seaplane flights were made as the vessel *Bear* cruised eastwards from Little America. These flights explored the Walgreen Coast, Thurston Peninsula, and Seraph Bay. From West Base itself, flights visited the

interior of Marie Byrd Land (four flights) and the Ruppert Coast, as well as extending to the Queen Maud Range by way of the Beardmore Glacier. From East Base (on Neny Island in Marguerite Bay), flights were made across the Antarctic Peninsula and Alexander I Island and along the length of George VI Sound and the Bowman Coast. To assist flight operations, a weather reporting station was established on the plateau to the east of the base.

11. Siple, 67.

12. West Base officially closed on 1 February, and East Base on 22 March 1941.

13. Siple, 74–75.

14. Siple, 76–81.

15. Siple, 64.

16. Rodgers, 144–45. The ritual was undoubtedly influenced by Byrd's experience in the Masons. Every so often he paused and asked if he should still proceed. Getting "yes" answers every time, Byrd administered the oath of the secret brotherhood.

17. Sullivan, 220.

18. Siple, 119.

19. *Time*, 31 December 1956.

20. Bertrand, 413, 423.

CHAPTER 7

Quotations where referenced are taken from Davis, J.K., *High Latitude*, and Siple, P., *90° South* (see Bibliography). Other material is taken from the author's notes.

1. Davis, 78–79. Captain Davis, after a life of fifty years as a mariner, retired from the post of Commonwealth Director of Navigation in 1949.

2. Iceberg bulletins (2000–2001) of the Antarctic Meteorological Centre (AMRC), University of Wisconsin.

3. Discussions and correspondence with Dr. Harry Keys.

4. Notes on the life of Father Daniel Linehan, S.J., from Boston College, Chestnut Hills, MA. A noted seismologist, Father Linehan served with Operation Deep Freeze from 1954–1958.

5. Quark Expedition news release.

6. A member of Scott's 1901 and 1910 expeditions, Dr. E. A. (Bill) Wilson died on the Barrier beside Scott and Bowers in March 1912 while returning from the South Pole.

Acknowledgements

We pay tribute to Captains Petr Golikov and Gennadiy Antokhin and their officers and crew, who with all the traditional skills of Russian seamanship and ice navigation took those trusting souls embarked upon *Kapitan Khlebnikov* safely across the wild waves of the Southern Ocean and through the Ross Sea pack on a memorable voyage to the end of the world.

Likewise we salute Chief Pilot Sergey Sirak and his fellow helicopter birdmen of Vladivostok Air, who spirited us securely to icy places that normally lie beyond the reach of mere mortals. Thanks also to Werner Stambach and Susan Adie and their team from Quark Expeditions; unruffled by frequently hostile elements, they directed the schedule of exploration and landings until we reached our goal at the Bay of Whales. Among their number we were fortunate to have as historian and raconteur

Bob Headland, Archivist of the Scott Polar Research Institute. Thanks again to the New Zealand team. Especially to Bob McCallum, New Zealand Government representative who opened the doors (literally) of the historic huts on McMurdo Sound, and Harry Keys, glaciologist, geologist, and vulcanologist, for his invaluable analysis of the giant 'bergs that flanked our course along the cliffs of the Great Ice Barrier.

The images from the voyage are principally by Catherine, a stalwart wife ever prepared to brave below-zero temperatures in the best traditions of polar photography. Stewart Campbell of Adventure Associates kindly made available his spectacular images of *Kapitan Khlebnikov* negotiating the Ross Sea pack, and Sue Jessop provided her close-up shots of icebergs, penguins, and the Weddell seal; the primitive sketches at the start of each chapter are from the author's own hand, with a list of photo credits opposite.

Thanks to Richard Hughes for his generous sub-editing, and Julia Burke for indexing assistance. Once again manuscript processing is in the capable hands of Paddy Elworthy, and Kerever Park has been unstinting in the provision of office services.

Finally, did we reach Farthest South?

Though Amundsen's vessel *Fram* may have swept a few minutes deeper into the Bay of Whales on its departing voyage back in 1911 (see table on page 176), when we watched *Kapitan Khlebnikov* dig its bow into the Bay's icy edge, the claim indeed could be made for Farthest South a ship had ever docked.

DAVID BURKE,
BURRADOO 2002

Photo Credits

All photographs taken on the voyage of *Kapitan Khlebnikov* are by Catherine Burke unless otherwise identified. Small sketches at the beginning of each chapter are the author's primitive art, sketched on the *Khlebnikov* voyage.

Australian Post Office: 24, 53, 85, 107 (lower)

Author's collection: 9, 12, 19, 21, 30, 34, 63, 71, 96–97, 115 (upper), 112, 123

Stewart Campbell, Adventure Associates: iv–v, xx, 12–13

Canterbury Museum (NZ): 28, 29

Canterbury Museum Collections: W. Colbeck photograph, Bernacchi collection, 3; Quartermain collection, 46; R.J. Orbell photograph, 50; O'Leary collection, 53; Kinsey collection, 149

John Fairfax: 51, 61, 78, 87, 93, 130–131

Val Jessop: 48–49, 76–77

National Archives & Records Administration, Washington D.C.: 94–95, 104–105, 107 (top), 124–125, 127, 133

National Geographic: 42–43, 121 (upper), 138–139

National Library of Australia T421, *Erebus and Terror* oil on canvas; 59.8 x 72.8 cm, Rex Nan Kivell Collection NK 2126: 27

National Library of Australia: 2, 4, 25, 45, 75, 101, 109, 110, 111

Royal Geographical Society: 72

Smithsonian Institution: 81, 115 (upper)

Tasmanian Museum & Art Gallery: 39

U.S. Army: 121 (lower)

U.S. Navy: x–xi, 40, 120 (upper), 142–143

Maps: K. Gilroy, J. R. Newland

Bibliography

Adams, H. *Beyond the Barrier with Byrd*. New York, M. A. Donohue & Co. , 1932.

Amundsen, R. *The South Pole*. London, John Murray, 1912.

Amundsen, R. My *Life as a Polar Explorer*. London, William Heinemann, 1927.

Balchen, B. *Come North With Me*. New York, E. P. Dutton & Co. , Inc. , 1958.

Bertrand, K. J. *Americans in Antarctica: 1775–1948*. New York, American Geographical Society, 1971.

Borchgrevink, C. *First on the Antarctic Continent*. London, George Newnes, Limited, 1901.

Burke, D. *Moments of Terror: The Story of Antarctic Aviation*. Sydney, New South Wales University Press, 1994.

Byrd, R. *Little America: Aerial Exploration in the Antarctic, the Flight to the South Pole*. New York, G. P. Putnam's Sons, 1930.

Byrd, R. Discovery: *The Story of the Second Byrd Antarctic Expedition*. New York, G. P. Putnam's Sons, 1935.

Davis, J. K. *High Latitude*. Melbourne, Melbourne University Press, 1962.

Debenham, F. *Antarctic: The Story of a Continent*. London, Herbert Jenkins, 1959.
Dufek, G. *Operation Deep Freeze*. New York, Harcourt, Brace & Co. , 1957.
Dufek, G. *Through the Frozen Frontier: The Exploration of Antarctica*. Leicester, Brockhampton Press, 1960.
Ellsworth, L. *Beyond Horizons*. New York, The Book League of America Inc. , 1938.
Grierson, J. *Challenge to the Poles*. London, Foulis, 1964.
Grierson, J. *Sir Hubert Wilkins—Enigma of Exploration*. London, Robert Hale Limited, 1960.
Hatherton, T. (ed.) *Antarctica*. London, Methuen, 1965.
Headland, R. K. *Chronological List of Antarctic Expeditions and Related Historical Events*. Cambridge, Scott Polar Research Institute, Cambridge University Press, 1989.
Hill, L. C. *Ellsworth Antarctic Relief Expedition, Report to Commonwealth Parliament*. Canberra Government Printer, 1936.
Huntford, R. *The Last Place on Earth*. London, Pan Books, 1985.
Huntford, R. (ed.) *The Amundsen Photographs*. London, Hodder & Stroughton, 1987.
Joerg, W. L. *Brief History of Polar Exploration Since the Introduction of Flying*. New York, 1930.
Martin, S. A. *History of Antarctica*. Sydney, State Library of New South Wales Press, 1996.
Mawson, P. *Mawson of the Antarctic: The life of Sir Douglas Mawson*. London, Longmans, Green & Co., 1964.
McGonigal, D. and Woodworth, L. *Antarctica—The Complete Story*. Melbourne, The Five Mile Press 2001.
Montague, R. *Oceans, Poles and Airmen*. New York, Random House, 1971.
Reader's Digest Services. *Antarctica: Great Stories from the Frozen Continent*. Surry Hills, Reader's Digest Services, 1985.
Rodgers, E. *Beyond the Barrier with Byrd*. Annapolis, Naval Institute Press, 1990.
Ronne, F. *Antarctic Conquest*. New York, G. P. Putnam's Sons, 1949.
Ronne, F. *Antarctic Command*. New York, The Bobbs Merrill Co. , 1961.
Ross, J. C. A *Voyage of Discovery and Research in the Southern and Antarctic Region*. London, John Murray, 1847.

Ross, M. J. *Ross in the Antarctic*. Whitby, Caedmon, Press, 1982.
Scott, R. F. *The Voyage of the Discovery*. London, Macmillan & Co. , 1905.
Shackleton, E. *The Heart of the Antarctic*. London, William Heinemann, 1910.
Siple, P. 90° *South: The Story of the American South Pole Conquest*. New York, G. P. Putnam's Sons, 1959.
Sullivan, W. *Quest for a Continent*. London, Secker & Warburg, 1957.
Swithinbank, C. *An Alien in Antarctica*. Blacksburg, The McDonald & Woodward Company, 1997.
Taylor, G. *Antarctic Adventure and Research*. New York, D. Appleton & Co. , 1930.
Thomas, L. *Sir Hubert Wilkins: His World of Adventure*. New York, McGraw-Hill, 1961.
Vaughan, N. D. *With Byrd at the Bottom of the World*. Harrisburg, Stackpole Books, 1990.

Index

METRIC CONVERSION

1 inch = 25.4mm

1 foot = 304.8mm

1 nautical mile = 1.15 mile

1 mile = 1.609 km

1 square foot = 0.093 square meters

1 ton = 1.016 metric tons

1 gallon = 4.55 liters

ALSO BY DAVID BURKE

Australia's Last Giants of Steam
Changing Trains
Come Midnight Monday
Darknight
Dreaming of Resurrection: A Reconciliation Story
Great Scott! (musical)
Great Steam Trains of Australia
Juggernaut
Julian: A Man Condemned? (dramatization)
Kings of the Iron Horse
Life of Mary Ward (video)
Making the Railways
Man of Steam
Moments of Terror: The Story of Antarctic Aviation
Monday at McMurdo
Railways of Australia
Road Through the Wilderness
The Observer's Book of Steam Locomotives
The World of Betsey Throsby
With Iron Rails